Genomic Negligence

Advances in genetic technology will lead to novel legal challenges. This book identifies four potential genomic claims which may be articulated as novel negligence challenges. Each of these claims is considered from the perspective of the English courts' approach to novel kinds of damage. It is argued that these novel genomic claims are unlikely to be favourably received given the current judicial attitude to new forms of damage. However, Victoria Chico argues that the genomic claims could be conceived of as harm because they concern interferences with autonomy. Each claim is considered from the perspective of a hypothetical English negligence system imbued with explicit recognition of the interest in autonomy. Chico examines how recognition of this new form of damage would lead to novel genomic negligence claims being treated in a way which they would not otherwise be if considered within traditional parameters of harm in negligence.

Victoria Chico is a Lecturer in Law at the University of Sheffield.

Biomedical Law and Ethics Library
Series Editor: Sheila A. M. McLean

Scientific and clinical advances, social and political developments and the impact of healthcare on our lives raise profound ethical and legal questions. Medical law and ethics have become central to our understanding of these problems, and are important tools for the analysis and resolution of problems – real or imagined.

In this series, scholars at the forefront of biomedical law and ethics contribute to the debates in this area, with accessible, thought-provoking, and sometimes controversial ideas. Each book in the series develops an independent hypothesis and argues cogently for a particular position. One of the major contributions of this series is the extent to which both law and ethics are utilised in the content of the books, and the shape of the series itself.

The books in this series are analytical, with a key target audience of lawyers, doctors, nurses, and the intelligent lay public.

Available titles:

Human Fertilisation and Embryology (2006)
Reproducing regulation
Kirsty Horsey & Hazel Biggs

Intention and Causation in Medical Non-Killing (2006)
The impact of criminal law concepts on euthanasia and assisted suicide
Glenys Williams

Impairment and Disability (2007)
Law and ethics at the beginning and end of life
Sheila McLean & Laura Williamson

Bioethics and the Humanities (2007)
Attitudes and perceptions
Robin Downie & Jane Macnaughton

Bioethics
Methods, theories, scopes
Marcus Düwell

Birth, Harm and the Role of Distributive Justice
Burdens, blessings, need and desert
Alasdair Maclean

Health Professionals and the Emergence of Distrust
Maladies of medical law
Mark Henaghan

Medicine and Law at the Limits of Life
Clinical ethics in action
Richard Huxtable

The Jurisprudence of Pregnancy
Concepts of conflict, persons and property
Mary Ford

About the Series Editor

Professor Sheila McLean is International Bar Association Professor of Law and Ethics in Medicine and Director of the Institute of Law and Ethics in Medicine at the University of Glasgow.

Genomic Negligence

An interest in autonomy as the
basis for novel negligence claims
generated by genetic technology

Victoria Chico

Routledge·Cavendish
Taylor & Francis Group

LONDON AND NEW YORK

First published 2011
by Routledge-Cavendish
2 Park Square, Milton Park, Abingdon, Oxon, OX14 4RN

Simultaneously published in the USA and Canada
by Routledge-Cavendish
711 Third Avenue, New York, NY 10017

*Routledge-Cavendish is an imprint of the Taylor & Francis Group, an
informa business*

British Library Cataloguing in Publication Data
A catalogue record for this book is available from the British Library

Library of Congress Cataloguing-in-Publication Data
Chico, Victoria.
  Genomic negligence: an interest in autonomy as the basis for
  novel negligence claims generated by genetic technology /
  Victoria Chico.
    p. cm.
  Includes bibliographical references and index.
  1. Human genetics–Law and legislation–England. 2. Medical
  personnel–Malpractice–England. I. Title.
  KD3410.G45.C48 2011
  344.4204'196–dc22   2011009769

ISBN: 978-0-415-49518-9 (hbk)
ISBN: 978-0-203-81781-0 (ebk)

Typeset in Garamond
by Wearset Ltd, Boldon, Tyne and Wear

Printed and bound in Great Britain by
CPI Antony Rowe, Chippenham, Wiltshire

Contents

Table of cases and statutes

EUROPEAN AND INTERNATIONAL CONVENTIONS

UK BILLS

UK CODES OF PRACTICE

Acknowledgements

This book began its life in 2003 as a PhD thesis under the attentive supervision of Professor Roger Brownsword and Mrs Norma Hird. It was examined in 2006 by Professor Vivian Harpwood and Professor Stuart Toddington, and the engaging viva provided me with many ideas which have been developed in this book. I would also like to thank the three reviewers for Routledge Cavendish who gave extensive and very helpful comments.

Whilst reworking the thesis for publication as a monograph, I have been lucky enough to benefit from the support and invaluable intellectual commentary of colleagues at Sheffield University. I am particularly grateful to Professor Derek Morgan who read and reread the manuscript, providing particularly insightful critical comments. I would also like to thank Dr Tawhida Ahmed, Dr Dimitrios Kyritsis, Professor Aurora Plomer and Dr Mark Taylor. I am indebted to these four colleagues who, after reading the manuscript, made invaluable comments and engaged with me in interesting and extensive discussions which have made this book better than it would otherwise have been.

I would also like to thank the members of the Jurisprudence Reading Group at the University of Sheffield who read and debated the autonomy chapter of this book in 2009.

I am particularly grateful to Dr John Coggon, Professor Emily Jackson and Dr Nicolette Priaulx, all of whom responded to out of the blue requests to read chapters of the manuscript and dedicated valuable time to providing me with particularly pertinent and detailed comments. I feel the book is much better for having considered their thoughts and arguments.

I am grateful for the diligent work of two research assistants; Anna Hescott and Anna Walsh, whose work was funded by the University of Sheffield School of Law Research Fund and the Sheffield Institute of Biotechnological Law and Ethics.

Finally, I want to thank Tim and my parents for being who they are and doing what they do, this book is dedicated to them.

Introduction

Modern genetic science enables the identification, manipulation and control of our genes. Genetic services arising from this science provide the means to predict and influence health and wellbeing. The ability to rapidly, increasingly and inexpensively sequence a person's entire human genome[1] means that people can discover if they possess genetic traits which predispose them to ill health. Knowing this information could alter the individual's perspective of her life and her feelings about herself in ways which do not necessarily change her choices, actions and desires. On the other hand, having information about the trajectory of her health may influence the individual to follow a different life plan to that which she might have followed if she had not had the information. The information might be useful in making decisions about early drug therapy, or risk avoidance measures, or about making lifestyle decisions which are not related to obviating or minimizing health risks. Knowing about personal genetic risks necessarily means that the individual also knows about risks to her relatives and to her future offspring. The ability to identify deleterious genetic traits at the embryonic stage and discard embryos that possess that genetic make-up means that information concerning one's own genetic traits has implications in reproductive decision-making.

The central idea in this book is that the ability to identify, manipulate and control our genes might lead to the development of new kinds of grievance that might present novel legal challenges. Four potential novel claims are identified which might arise from genetic services: two concerning reproductive genetic services and two concerning genetic information. For the most part, these hypothetical genomic claims arise due to the culpable carelessness of an individual who has undertaken to assist the aggrieved party in her genetic project. In the absence of dedicated regulation, or a contract, it is argued that these genomic claims are most likely to be articulated as novel

---

1 For information on the development of a new generation of sequencing technologies which enables sequencing DNA at unprecedented speed, promoting novel biological applications see S. C. Schuster, 'Next-Generation Sequencing Transforms Today's Biology' 5 (2008), *Nature Methods* 16.

negligence challenges. Thus, it is not suggested here that there should be a new tort solely aimed at providing redress for claims which arise from genetic technology.[2] Here the potential claims are, in the first instance, considered from the perspective of existing English negligence principles.

English negligence law experiences periods of expansion and restriction. Generally, following an expansive period in the 1970s and 1980s, the law is currently experiencing a restrictive period.[3] The present attitude to the imposition of new duties of care is particularly restrictive. Rather than analysing the novel genomic claims from the perspective of English negligence principles generally, this book considers the particular legal issues which are likely to be of greatest significance to the question of the recognition of the novel genomic negligence claims. Thus, the focus is on the legal concept of duty of care because of its use as a mechanism for declining to recognize new interests within the tort of negligence. This is not to suggest that the genomic negligence claims will not present issues with respect to breach, causation and remoteness of damage, but consideration of these issues will have to be left to another occasion. In the context of this duty of care focus, the discussion concentrates on the particular interests which are perceived to have been interfered with in each novel claim, and considers these within the context of English negligence law's current approach to recognition of new interests.

It is widely recognized that damage is the gist of the action in negligence;[4] hence the focus on damage here where the setback to the interest upon which the claim is based does not fit within the range of harms traditionally recognized by English negligence law. In essence, the grievances which arise in the novel genomic negligence claims concern unwanted birth, the failure to secure the birth of the desired child, the failure to give relevant genetic information or the unsolicited disclosure of unwanted genetic information. These grievances do not fit within the traditional scope of damage in negligence; largely physical harm to person or property, and in some exceptional circumstances psychiatric harm or a setback to pure financial interests. In each of these claims it might be argued that the claimant could present the setback to her interests as the loss of an opportunity of some sort. However, the House of Lords has not been particularly favourable

---

2   Although for the consideration of an approach which recognizes the possibility of a new blockbuster tort based on human dignity as a way of responding to novel genomic claims see R. Brownsword, 'An Interest in Human Dignity as the Basis for Genomic Torts' 42 (2003), *Washburn Law Journal* 143.

3   Although there are some exceptional areas where the House of Lords appears to be in favour of expanding liability in negligence. See the approach to causation in *Fairchild v Glenhaven Funeral Services* [2003] 1 AC 32. See also the liability of local authorities for welfare and educational issues in *Phelps v Hillingdon London Borough Council* [2001] 2 AC 619 and *Barrett v Enfield London Borough Council* [2001] AC 550.

4   J. Stapleton, 'The Gist of Negligence: Part I: Minimum Actionable Damage' 104 (1988), *Law Quarterly Review* 213.

to the reformulation of damage as lost chances in recent years.[5] Thus, after considering each novel genomic claim within the context of existing analogous English cases,[6] this book argues that the genomic claims could be conceived as interferences with the aggrieved party's autonomy.

Some might contend that a system that provides redress for the interference with autonomy is based on rights vindication[7] as opposed to recognizing and remedying harm. From the rights vindication perspective, any recognition of the interest in autonomy would be a tall order for a negligence system which is premised on damage being the gist of the action. To a certain extent, the tort of battery is able to recognize interferences with autonomy in a way that might be seen simply as vindicating rights because the tort is actionable per se.[8] However, the question of whether an interference with autonomy breaches rights or occasions harm depends on one's perspective; some prominent commentators argue that interference with autonomy can be recognized as a harm.[9] On the basis that interference with an individual's autonomy can be conceived as occasioning harm to that individual, each of the novel claims is considered from the angle of a hypothetical English negligence system which is imbued with explicit recognition of the interest in autonomy. It is argued that the recognition of this new form of damage would offer a real possibility of the novel genomic negligence claims being brought forward and given genuine consideration by the courts in a way which they might not be if the claims were considered within the traditional parameters of harm in negligence. This book does not seek to argue that the explicit recognition of interference with autonomy as damage would be the best way for English negligence law to respond to these genomic claims, this issue requires further discussion.[10] However, it works from the basis that one

---

5   Particularly with regard to lost chances of avoiding physical harm. See *Hotson v East Berkshire Health Authority* [1987] AC 750 and *Gregg v Scott* [2005] 2 AC 176. However, see the discussion of the more favourable approach to lost chances of economic gain in Chapter 6.

6   With regard to some of the genomic claims, cases from other jurisdictions are considered where there is particularly persuasive or well-developed authority. However, the focus of this book is English negligence law. It does not provide a comparative analysis or comprehensive discussion of the law in other jurisdictions. For a book which does provide comparative perspectives in the context of reproductive claims (to which the former two claims considered in this book pertain) with the USA, the Commonwealth and Europe see J. K. Mason, *The Troubled Pregnancy: Legal Wrongs and Rights in Reproduction,* Cambridge: Cambridge University Press (2007).

7   Generally this book does not employ the term 'right' and the analysis herein is not a rights-based analysis.

8   For a consideration of the novel claims within the context of the tort of battery see Chapter 2.

9   See, in particular, D. Nolan, 'New Forms of Damage in Negligence' 70 (2007), *Modern Law Review* 59, 79–80; N. Priaulx, *The Harm Paradox: Tort law and the Unwanted Child in an Era of Choice,* London: Routledge (2007); E. Jackson, 'Informed Consent to Medical Treatment and the Impotence of Tort' in S. McLean (ed.) *First Do No Harm,* Aldershot: Ashgate (2006), 273, 274.

10 However, the impetus for this argument is provided by two fairly recent House of Lords decisions which arguably recognize the interest that the individual has in not having her autonomy interfered with by the negligence of others. See *Rees v Darlington Memorial Hospital NHS Trust* [2004] 1 AC 309 and *Chester v Afshar* [2005] 1 AC 134.

of the essential features of the common law, premised on a doctrine of judicial precedent, is to maintain consistency in the law.[11] Thus, it is argued that by recognizing the harm occasioned by interference with autonomy, the courts could respond comprehensively and consistently to grievances that might arise from negligence in the provision of genetic services.

Given that the concept of autonomy forms the heart of the discussion of how English negligence law might respond to the novel genomic negligence claims, this book makes suggestions about the conception of autonomy that the courts could adopt. Autonomy is not a unified principle; it is subject to many different conceptions, so much so that probably the only things that hold constant between theories of autonomy is that it is a feature of humans, and that respect for autonomy is generally a good thing.[12] The analysis of autonomy herein starts from the position that the concept of autonomy explicitly adopted by the English tort system, within the context of consent in the tort of battery, amounts to a content-neutral individualistic interpretation of the concept. However, although this is the approach explicitly adopted, it is argued that in reality the courts actually require something more substantive before they will deem that the patient's decision is autonomous and, therefore, ought to be protected by the court on the basis of an appeal to autonomy.

The question of whether autonomy is a content-neutral or a substantive/value-based concept has been debated by philosophers for many years. Where autonomous ends need not be based on particular values or substance, the question of whether a choice, action or desire is autonomous may depend on whether the arrival at that choice, action or desire fulfilled particular

---

11   The doctrine of precedent is premised on promoting consistency. One of the earliest acknowledgements of this can be found in *Mirehouse v Rennell*, 1 Clark & Finnelly 527, Parke J, 546. 'Our common law system consists of applying to new combinations of circumstances those rules of law which we derive from legal principles and judicial precedent; and for the sake of attaining uniformity, consistency and certainty, we must apply those rules, where they are not plainly unreasonable and inconvenient, to all cases which arise; and we are not at liberty to reject them, and to abandon all analogy to them, in those to which they have not been judicially applied, because we think that the rules are not as convenient and reasonable as we ourselves could have devised'. In recognizing the importance of certainty and consistency in the law, Parke J's dictum was restated in *R v Simpson* [2004] QB 118, Lord Woolf CJ, 128. For a recent example of the importance of maintaining consistency in the law, which is particularly relevant in this book, see *Rees v Darlington Memorial Hospital NHS Trust* [2004] 1 AC 309. In *Rees* the majority was keen to award a lump conventional sum in all cases to avoid the anomalies that might arise where damages are not awarded for the costs of raising a healthy child, but awards are made in other cases on a discretionary basis. See also *Rees v Darlington Memorial Hospital NHS Trust* [2004] 1 AC 309, Lord Bingham, 317; Lord Nicholls, 319; Lord Millett, 349, all relying on the dicta of Waller LJ in the Court of Appeal.

12   Unless it interferes with other more important legitimate interests of others. Some authors argue that conceptions of individual autonomy cannot provide a sufficient and convincing starting point for ethics within medical practice. See, in particular, C. Foster, *Choosing Life, Choosing Death: The Tyranny of Autonomy in Medical Ethics and Law*, Oxford: Hart (2009); O. O'Neill, *Autonomy and Trust in Bioethics*, Cambridge: Cambridge University Press (2002); G. Laurie, *Genetic Privacy, A Challenge to Medico-Legal Norms*, Cambridge: Cambridge University Press (2002); and G. M. Stirrat and R. Gill, 'Autonomy in Medical Ethics After O'Neill' 31 (2005), *Journal of Medical Ethics* 127.

procedural criteria. This book employs the concept of autonomy as an evaluative tool to consider how English negligence law might react to the novel genomic claims. It does not seek to make an argument for any particular theory of autonomy in abstract. However, it argues that a conception of autonomy which has at least some substantive element might function better as a legal principle than a liberal content-neutral interpretation, because of its ability to enable some objective analytical purchase about what the principle of autonomy actually consists of.[13] On this basis, this book discusses how the principle of autonomy might be interpreted as a principle that is underpinned by the concept of rationality. First, it concentrates on rationality as a substantive concept or one denoting particular value which would inject particular content into the principle of autonomy. Following this, the focus is on rational procedure as a means of explaining the autonomy of choices, actions and desires. If the law is to recognize a novel action, it needs to consider that the claim is meritorious and, therefore, worthy of legal protection. A principle of autonomy which is premised on particular substance might be preferable as a legal notion[14] because it provides content which one can at least attempt to justify as worthy of admiration.[15] Despite not personally holding to such a substantive interpretation of autonomy, Gerald Dworkin notes the practical benefits of this approach over 'a conception of autonomy that has no particular content, that emphasizes self-definition in abstraction from the self that is so defined, (which) seems too thin, too formal to be of much value'.[16]

The substantively[17] and procedurally[18] rational conceptions of autonomy are considered in abstract in Chapter 3. Following this, the conceptions of autonomy are considered in the context of the particular issues raised in relation to each of the individual genomic claims to evaluate how a system of negligence law, which is imbued with a recognition of interference with autonomy as harm, might respond to that claim. In this way, this book considers how English negligence law might provide comprehensive and consistent protection when an individual's autonomy interests are interfered with as a result of genomic negligence.

---

13 The English courts are wary of employing principles with respect to which they can foresee 'serious definitional difficulties and conceptual problems in the judicial development' thereof. See Mummery LJ in *Wainwright v Home Office* [2002] QB 1334, 1351.

14 As opposed to a liberal notion of autonomy which is largely unconditional.

15 G. Dworkin, *The Theory and Practice of Autonomy,* Cambridge: Cambridge University Press (1988), 30.

16 Ibid.

17 The major problem with a conception of autonomy as imbued with rationality, which is premised on particular values, is that the question of what is valuable raises a legitimacy problem where there is no clear idea of what value consists of. This book takes a legal approach to how the question of whether there is value in claims seeking to establish novel kinds of damage might be determined. See the discussion in Chapter 3.

18 The procedural account of autonomy adopted as an evaluative tool here largely rests on the hierarchical account of autonomy adopted by Gerald Dworkin and Harry Frankfurt.

Chapter 1

Some genetic science which is of significance to novel genomic negligence claims

Introduction

Modern genetic capabilities are changing the way we think about health by enhancing understanding of how genes function and enabling the prediction and modification of biological futures. Since everybody carries around three to nine deleterious or diseased genes, which put them at risk of something, modern genetic technology is potentially relevant to everyone.[1] In 1992 a team of scientists began a research project; the Human Genome Project (HGP), which planned to map and sequence the entire human genome. The HGP involves the discovery and sequencing of the DNA in a single human cell. Its primary goal is the listing and location of our *genes*.[2] The information derived from the completed project is beginning to provide clues about how to combat genetic diseases.[3]

The ultimate aim of the HGP is to facilitate the diagnosis, understanding and treatment of the more than 5,000 genetic conditions that affect mankind.[4] There have been significant developments in terms of testing for

---

1 S. M. Suter, 'Whose Genes Are These Anyway? Familial Conflicts over Access to Genetic Information', 91 (1993) *Michigan Law Review*, 1854, 1858. Current estimates state that 60 per cent of the people living in the UK are likely to develop a disease that is at least partially genetically determined by the age of 60. Also, approximately one in 30 children is born with a genetic condition: M. Urner, 'Genetic Interest Group' Changes Name to 'Genetic Alliance UK', *Bionews* 560 (June 2010).

2 On 14th April 2003, the International Human Genome Consortium announced successful completion of the project: International Human Genome Sequencing Consortium, 'Finishing the Euchromatic Sequence of the Human Genome', 431 (2004) *Nature*, 931.

3 Thus, the information derived from the project does not lead directly to treatment for genetic disease, rather it provides the foundation for researching genetic disease. It does, however, enable the sequencing of one's personal genome and the information derived from this personal sequencing might be used to make predictions about personal disease risks. New and faster methods of sequencing the genome known as 'next-generation sequencing' can provide access to full genome sequencing so that the entire genetic code of an individual can be deduced all at once. See S. C. Schuster, 'Next-Generation Sequencing Transforms Today's Biology', 5 (2008) *Nature Methods*, 16.

4 For general comment on the Project and its aims see, J. D. Watson, 'The Human Genome Project: Past Present and Future', 248 (1990) *Science*, 44.

genetic conditions, however, little progress has been made with respect to treatment. Now the human genome has been mapped, a major focus for genetic research is on translating the information gained from the HGP into tangible health benefits. As means of ameliorating genetic disease are discovered, more people will seek testing and treatment, which may present new legal challenges.[5]

The aim of this chapter is to provide background information on the genetic capabilities, which is relevant to the novel legal claims whose discussion forms the core of this book. It is hoped that the provision of this information will facilitate the understanding of how the novel claims might arise. The claims fall into two broad categories;[6] claims arising from embryo screening[7] and claims arising from genetic information.[8]

This chapter is split into two parts. Part I focuses on embryo screening. After a brief consideration of the transmission of hereditary genetic disorders, this part focuses on how the technique of preimplantation genetic diagnosis (PGD) can avoid the birth of a child with a particular genetic defect.

Part II focuses on the generation of genetic information, concentrating on the discrepancy between the availability of testing for, and treatment of, genetic disease. This leads into a discussion of the distinction between disorders which are caused by a single gene defect and those which are caused by the interaction of genes and the environment.[9]

Embryo testing

People inherit two copies of every gene, one from their mother and one from their father. A disorder carried by genes can be either a dominant or a recessive trait. A genetic trait is considered dominant if it is expressed in a person who has only one copy of that gene. Thus, if the affected gene is dominant, a person with one or two copies of the gene will have the disorder. A recessive trait is expressed when two copies of the gene carrying the disorder are present. Thus, if the affected gene is recessive, only a person with two copies of the gene will have the disorder. If two healthy carriers of a defective, recessive gene reproduce, there is a 25 per cent chance that their offspring will inherit two affected genes and suffer the disorder, a 50 per cent chance that the child will inherit one affected gene and be a carrier, and a 25 per

---

5 The potential novel legal implications of the Human Genome Project have been recognized, with 3 per cent of the project's research budget being devoted to the ethical and legal ramifications of mapping the genome.

6 Given the purpose of this chapter there are many things that are not said about the genetic technology in relation to these two broad categories.

7 The claims considered in Chapters 4 and 5.

8 The claims considered in Chapters 6 and 7.

9 Thus, this chapter is not a comprehensive account of modern human genetics.

cent chance that the child will not inherit a faulty gene.[10] However, dominant genetic disorders occur when only one copy of the faulty gene is inherited. With dominant disorders there is a 50 per cent chance that a child will be affected by the disorder if one parent carries the faulty gene and a 75 per cent chance if both parents do.[11] Around 50 per cent of dominant disorders are late onset,[12] thus an individual might have reproduced before learning of her own genetic susceptibility.

Potential parents who risk passing on a genetic disorder to their offspring in one of the ways described above, may want to avoid having a child with a serious genetic condition; Pre-implantation Genetic Diagnosis (PGD) gives them the opportunity to do this. PGD involves creating a number of embryos via *in vitro* fertilisation (IVF); one or two cells are then taken from those embryos at the eight-cell stage[13] to determine whether the embryo is affected by the genetic disorder. An unaffected embryo is transferred to the uterus and allowed to develop to term. The use of this technique to screen out embryos with severe and fatal disorders now appears to be accepted and its scope is being steadily expanded.[14] Embryo testing is regulated by the Human Fertilisation and Embryology Act 1990.[15] The Human Fertilisation and Embryology Authority (the Authority) continues to issue licences for the specific conditions that can be screened out via PGD. It maintains a central list of disorders for which PGD is currently licensed on the HFEA website. There are presently 139 genetic disorders on the list.[16]

Since the 1990s, the Authority has permitted PGD for couples at risk of passing on genes which virtually guarantee that the resulting individual will manifest a particular genetic condition.[17] The penetrance of the particular gene is an important factor in deciding whether to allow PGD. Penetrance describes

---

10 Where only one parent has the recessive faulty gene, the healthy gene of the other parent is sufficient to override it.

11 It is very rare for a person to inherit two copies of a dominant gene carrying a genetic disorder, but where two copies of a dominant gene are inherited it will produce a much more serious form of the disorder.

12 G. T. Laurie, *Genetic Privacy A Challenge to Medico-Legal Norms,* Cambridge: Cambridge University Press (2002), 95.

13 The eight-cell stage is a period in embryonic development when the embryo has undergone three divisions from a single cell (one into two, two into four and then four into eight), resulting in eight cells. This usually occurs on about day three of the embryo's development.

14 Human Fertilisation and Embryology Authority, Press Release Archive, *Authority Decision on PGD Policy* (10 May 2006). The Authority agreed that it should consider the use of PGD embryo testing for conditions such as inherited breast, ovarian and bowel cancers which do not always manifest, and when they do it is not until later in life.

15 Schedule 2, s. 1 ZA.

16 Available HTTP: www.hfea.gov.uk/pgd-screening.html (accessed 11 May 2010). Before the Authority will permit testing for a particular genetic condition, its members must agree that the condition is sufficiently serious. There are 16 conditions on the list currently awaiting consideration by the HFEA. Available HTTP: www.hfea.gov.uk/5643.html (accessed 11 May 2010).

17 Examples are Huntington's disease, myotonic dystrophy (dominant), cystic fibrosis, spinal muscular atrophy (recessive), beta thalassaemia, sickle cell disease (haemoglobinopathies), Lesch-Nyhan disease, Duchenne muscular dystrophy and haemophilias (X-linked).

the degree to which individuals possessing a particular genetic mutation express the trait caused by that mutation. A highly penetrant gene will express itself regardless of the effects of environment, whereas a gene with low penetrance will not always produce the symptoms with which it is associated. Initially, the Authority only permitted PGD for disorders with a high degree of penetrance.[18] However, it has recently expanded its remit to allow screening for lower penetrance conditions. Following the response to the HFEA public consultation Choices and Boundaries,[19] HFEA members decided that couples should be allowed to have their embryos tested for genes which make carriers susceptible to inherited cancers, such as breast and ovarian cancer.[20] These cancers are associated BRCA1 genetic variants and the penetrance is 50–80 per cent.[21] Thus, up to 50 per cent of individuals who possess the gene variant will not go on to develop breast or ovarian cancer.

In addition to allowing testing for particular genetic disorders, the Authority provides licenses to determine the tissue type of certain embryos so that the resulting child might provide treatment for an existing sibling. In 2001 the Authority announced its decision to allow tissue typing in conjunction with PGD for serious genetic diseases, so that couples at risk of passing on a genetic disorder could select embryos free from the disease and with tissue matching that of the existing child, so that stem cells from the resulting baby's umbilical cord blood might be used to provide treatment for a sibling.[22] Initially, this technique was only permitted if the created 'saviour sibling' was at risk of suffering from the condition which the existing child suffered from. However, in July 2004 the Authority relaxed its policy to allow tissue typing in isolation to determine whether a particular embryo could act as a saviour to her

18 The penetrance of some single gene disorders, such as Huntington's disease, is virtually 100 per cent. See Human Genetics Commission, *Making Babies: Reproductive Decisions and Genetic Technologies* (January 2006), paragraph 4.17.

19 Human Fertilisation and Embryology Authority, *Choices and Boundaries Report: A Summary of Responses to the HFEA Public Discussion* (2006).

20 Human Fertilisation and Embryology Authority, Press Release Archive, *HFEA-Authority Decision on PGD Policy* (10 May 2006).

21 P. Brice, 'HFEA Consultation on PGD for Lower Penetrance Conditions' (3 October 2005), Public Health Genetics Unit. Available HTTP: www.phgu.org.uk/news/2017 (accessed 20 June 2007).

22 Human Fertilisation and Embryology Authority, Press Release Archive, *HFEA to Allow Tissue Typing in Conjunction with Preimplantation Genetic Diagnosis* (13 December 2001). Umbilical cord blood is blood that remains in the placenta and in the umbilical cord after childbirth. Cord blood is collected because it contains stem cells which can be used to treat genetic disorders. Stem cells are characterized by the ability to renew themselves and differentiate into a diverse range of specialized cell types. They are pre-differentiated and, therefore, different to cells which are already differentiated and cannot change into other types of cell. Stem cells derived from umbilical cord blood are highly plastic in that they can be encouraged to transform with characteristics consistent with many different types of cell. A stem cell transplant between tissue-matched siblings carries a greater than 90 per cent success rate: Communication with Professor Ajay Vora, professor of paediatric haematology and consultant paediatric haematologist (Professor Vora treated Charlie Whitaker after his parent's project to create a saviour sibling for him in Chicago was successful).

sibling.[23] In other words, the child to be created need not be at risk of the condition. Treatment to procure saviour siblings has now been put on a statutory footing via the 2008 amendments to the 1990 Act. Schedule 2, Section. 1 ZA (1) (d) of the Act states:

> in a case where a person ('the sibling') who is the child of the persons whose gametes are used to bring about the creation of the embryo (or of either of those persons) suffers from a serious medical condition which could be treated by umbilical cord blood stem cells, bone marrow or other tissue of any resulting child, establishing whether the tissue of any resulting child would be compatible with that of the sibling.

Thus, it is now clear that this treatment is available where the prospective child is not at any increased risk of being born with the particular genetic condition.

Genetic information

Altering the nuclear or mitochondrial DNA of a cell while it forms part of an embryo is prohibited by the 1990 Act.[24] Thus, upon discovery of a defective gene, an embryo can only be de-selected. However, with regard to existing persons with genetic diseases, a major research focus is on correcting or replacing defective genes. The latter two claims considered in this book concern novel grievances that might arise in the face of the increasing availability of genetic information. More than 900 genetic tests are currently available.[25] Progress in the treatment of these genetic disorders is much slower.

Genetic disorders are either monogenic or polygenic. Monogenic disorders occur due to a single gene which is passed on to subsequent generations via dominant, recessive and X-linked inheritance patterns.[26] Zimmern states that the relative predictive power of genetic testing depends on the condition being

---

23   See Human Fertilisation and Embryology Authority, Press Release Archive, *HFEA Agrees to Extend Policy on Tissue Typing* (21 July 2004).

24   Human Fertilisation and Embryology Act 1990, Schedule 2, Section 1 (4).

25   See Human Genome Project Information. Available HTTP: www.ornl.gov/sci/techresources/Human_Genome/home.shtml (accessed 1 July 2007).

26   Many disease traits are carried on the X chromosome. The X chromosome is a sex chromosome. With regard to sex chromosomes, females have two X chromosomes and males have one X and one Y. Inherited genetic disorders that are carried on the sex chromosomes are referred to as sex-linked disorders, and disorders which are carried on other chromosomes are referred to as autosomal. Sex chromosomal genetic disorders predominantly affect males because they carry only one X chromosome and there is no second copy of these genes on the Y chromosome. Therefore, a male has only one copy of these genes. If his copy is damaged or defective, he has no normal copy to override or mask the defective one. X-linked conditions are passed from mother to son because a son's X chromosome always comes from his mother. A son born to a female carrier has a fifty-fifty chance of having the disorder. X-linked disorders include Hemophilia A, Duchenne muscular dystrophy, and Lesch-Nyhan syndrome.

considered.[27] Directly obtained genetic information for high penetrance mono-genic disorders such as Huntington's disease will be highly predictive of the onset of that disease.[28] Furthermore, once a person has tested positive for a par-ticular genetic disorder the likelihood that her siblings or offspring will carry that disorder can be predicted with some accuracy. Dominant single-gene dis-orders, in particular, have a tendency not to manifest themselves until later in life.[29] Where a person tests positive for a monogenic, dominant genetic disorder there is a 50 per cent chance that her siblings will possess the gene if one of her parents possesses the gene, and a 75 per cent chance that they will possess the gene if both parents do. Furthermore, there is 50 per cent chance that each of her children also possesses the gene, rising to a 75 per cent chance if her repro-ductive partner is also affected.[30]

Relatively speaking, monogenic conditions are rare. A far greater number are affected by multifactorial or polygenic disorders, whereby a particular condition is caused by the effects of multiple genes or multiple genes in combination with lifestyle and environmental factors. Examples of multifac-torial disorders include cancer, heart disease, diabetes, hypertension, allergies and epilepsy. Determining a person's risk of manifesting a multifactorial dis-order is more difficult than determining her risk of manifesting a monogenic disorder, because although multifactorial disorders often cluster in families, they do not have a clear-cut (Mendelian) pattern of inheritance.[31] A person's genetic makeup merely makes her more susceptible to a particular genetic disorder; whether the condition manifests in that individual depends on the internal interaction of her genes or the interaction of her genes with the environment. Possession of the deleterious genetic trait is not highly predic-tive in the same way as a positive test for a monogenic disorder.

The predictive ability of genetic information and existence of treatment for the condition are likely to have some bearing on whether an individual wants to receive information about her genetic risks. She might want to

27 R. Zimmern, 'What is Genetic Information?' 1 (2001) *Genetics Law Monitor* 9, 13.

28 The Human Genetics Commission states that Huntington's disease is almost 100 per cent penetrant. See Human Genetics Commission, *Making Babies: Reproductive Decisions and Genetic Technologies* (January 2006), paragraph 4.17.

29 G.T. Laurie, *Genetic Privacy A Challenge to Medico-Legal Norms* (2002), Cambridge: Cambridge University Press, 95.

30 As explained above, it is very rare for a person to inherit two copies of a dominant gene carrying a genetic disorder, i.e. one from each parent. But if two copies of a dominant gene are inherited it will produce a much more serious form of the disorder.

31 Gregor Mendel first noted how the inheritance of traits might predictably occur in peas. These basic rules of inheritance are sometimes called Mendelian inheritance. Under Mendelian inheritance, a dominant trait is one that appears even when the second copy of the gene for that trait is different. For example, for the seeds of Mendel's peas, smooth is dominant over wrinkled. Thus, if a pea plant contains one gene for smooth and one for wrinkled, the seed will be smooth. Wrinkled is a recessive trait, which is one that only appears when two copies of it are present.

know if there is a 50 per cent chance that she possesses a genetic mutation, which will virtually guarantee that she will manifest a genetic disorder. On the other hand, if there is only a 5–10 per cent chance that she possesses a deleterious genetic makeup which only means she is more susceptible to a particular genetic disorder, she may not want to know. Of course, this does not necessarily mean that the greater the chance that the condition will manifest, the greater the likelihood that 'at risk' individuals will want to know. In the current genetic climate where very little can be done to cure, delay or treat genetic disorders, those who are highly likely to manifest a particular disorder may be the ones who desire ignorance the most.

Given the aetiology of multifactorial genetic conditions, it might be possible to significantly reduce or even eliminate an individual's risk of manifesting a particular genetic disorder by simply informing her of the environmental toxins which, when combined with her particular genetic makeup, increase her susceptibility to a particular condition. This is, of course, dependant on the identification of how specific genes and environmental toxins react. Although identifying the genes involved in multifactorial conditions might be more difficult than identifying those involved in monogenic conditions, scientists have become adept at discovering new disease-related genes. The challenge lies in discovering ways to compensate for those genes which are defective.[32] If multifactorial disorders can be prevented simply by avoiding environmental toxins or early drug therapy, they may be significantly easier to avoid than those disorders which would require replacement gene therapy.[33]

The concept of gene-environment interaction is becoming a central theme in epidemiologic studies that assess causes of human disease in populations.[34] Once concrete links are made between genes and the environment it may be easy, from a theoretical perspective, to avoid the relevant toxins.[35] The

32 Genome-wide association studies (GWAS) examine genetic variation across a given genome. These studies are designed to identify genetic associations with observable traits such as blood pressure or weight, or why some people get a disease or condition. See T. A. Pearson and T. A. Manolio, 'How to Interpret a Genome-Wide Association Study' 299 (2008) *Journal of the American Medical Association*, 1335. Genome-wide association studies involve comparing the genome of people with a particular disease to a control group in which the disease is not present. If genetic variations are more frequent in people with the disease, the variations are said to be associated with the disease.

33 In the future, monogenic disorders might be treated by other means. For example, by replacing the protein that would have been produced by the faulty gene. However, this does not amount to a cure.

34 M. J. Khoury, T. H. Beaty, B. H. Cohen, *Fundamentals of Genetic Epidemiology* (1993), Oxford: Oxford University Press. Targeted research is required to investigate what role the environment plays in asthma and other allergies, neuro-immune disorders and cancer, and how environmental risks are influenced by genetic factors. EUR 21458 – European Union Research on Environment and Health – *Expanding Knowledge to improve our Well-Being*, Luxembourg: Office for Official Publications of the European Communities (2005), 8.

35 Common non-genetic factors that influence the manifestation/progression of genetic diseases include diet, exercise, stress, alcohol, drugs and exposure to toxic chemicals or radiation. See ibid. 12 at 97.

research focus, in relation to monogenic conditions, is different; many single gene mutations have already been identified; the particular scientific hurdle lies in devising ways to replace those faulty genes. The major focus in this respect is gene therapy. In theory, gene therapy has the potential to treat or even cure genetic diseases. It involves the insertion of genes into an individual's cell to treat diseases by replacing deleterious mutant alleles[36] with functional ones, thereby attempting to correct the underlying problem by introducing healthy copies of the damaged or missing genes into some of the patient's cells.[37] Its promise has not escaped research commissioners. In the White Paper 'Our Inheritance, Our Future: Realising the Potential of Genetics in the NHS',[38] the director of research and development invited proposals for research on the translational mechanisms and service developments needed to move gene therapy for inherited single gene disorders on from a research environment to the clinical environment.[39]

One of the barriers to progress in the field of gene therapy is the lack of a safe vector.[40] A vector is a vehicle which is used to transport the required genetic material to the target cell. There is significant interest in modified viruses as vectors because viruses attack their hosts and introduce their genetic material into the host cell. Rigby notes that while the underlying principle of gene therapy is not particularly complicated, the practice is anything but simple. He suggests that the major problem in designing effective models of gene therapy is in designing vectors which are efficient and safe, and which can be injected into the body. Most vectors provoke immune responses in the patient, or are inactivated by components of human blood, or are very inefficient.[41] Thus, it is clear that if gene therapy is to become a standard way of treating people then there will have to be dramatic improvements in vector design.

Nevertheless, the Chairman of GTAC has noted that in some situations: 'gene therapy is beginning to result in real clinical promise'.[42] The Gene

---

36 An allele is a gene that is found in one of two or more different forms in the same position in a chromosome. Available HTTP: http://dictionary.cambridge.org/dictionary/british/allele (accessed 11 June 2010).

37 Available HTTP: www.dh.gov.uk/ab/GTAC/Genetherapy/index.htm (accessed 11 June 2010).

38 Department of Health (2003), *Our Inheritance, Our Future: Realising the Potential of Genetics in the NHS*, Cm 5791, London, Stationery Office.

39 Department of Health, *Gene Therapy Research Invitation to Tender* (2003).

40 The Human Genome Project Information website gene therapy page. Available HTTP: www.ornl. gov/sci/techresources/Human_Genome/medicine/genetherapy.shtml (accessed 16 July 2007) confirms that difficulty with viral vectors is one of the factors which has kept gene therapy from becoming an effective treatment for genetic disease.

41 P. Rigby, 'Gene Therapy: Simple in Theory but Difficult in Practice', *Mill Hill Essays* (1995), MRC National Institute for Medical Research.

42 Professor Norman Nevin, Gene Therapy Advisory Committee, Eleventh Annual Report (2004), 9.

Therapy Clinical Trials Worldwide website is provided by the Journal of Gene Medicine.[43] The website shows the number of approved, ongoing or completed clinical trials worldwide. Currently, the number of worldwide studies addressing monogenic genetic disorders account for 7.9 per cent of the whole. Studies addressing the multifactorial genetic conditions: cancer, cardiovascular and infectious diseases account for 64.5 per cent, 8.7 per cent and 8 per cent respectively. Currently, clinical research into gene therapy is at a very early stage; 60.3 per cent of studies are at Phase I, 18.9 per cent at Phase I/II and 26.3 per cent at Phase II. This means that only 4.3 per cent[44] of studies are at Phase II, III, or VI.[45] Thus, most treatment by way of gene therapy is still experimental, but the prevalence of studies worldwide[46] and the promising results of some early clinical trials[47] suggest that in the future gene therapy might provide effective treatments for those at risk of genetic disease.[48]

In the face of the increasing availability of genetic testing, individuals will have to make difficult decisions about whether to seek out such information. However, the nature of genetic information adds an even more difficult dimension to the issue of its control. Generally, it is thought that the control of information should be located in, and exercised by, the person to whom the information belongs or to whom it relates – that is the source of the information – should exercise that control.[49] However, in relation to genetic

43   Available HTTP: www.wiley.co.uk/genetherapy/clinical (accessed 12 March 2010).
44   0.1 per cent of studies only have a single subject which appears to exclude them from the phase data.
45   This amounts to just 70 studies out of 1579 worldwide. Available HTTP: www.wiley.co.uk/gene-therapy/clinical (accessed 12 March 2010).
46   A large number of clinical gene therapy trials continue to be approved worldwide. In 2008, 116 trials were approved. The only year a greater number was approved was 2006 when 117 were approved. Available HTTP: www.wiley.co.uk/genetherapy/clinical (accessed 12 March 2010).
47   In essence, Phase I clinical trials are small, recruiting up to 30 people to discover things such as dosage and side effects. Phase IV trials are done after a drug has been shown to work and has been granted a licence. These trials look at drugs that are already available for doctors to prescribe, rather than new drugs that are still being developed. The reasons for these trials are to discover more about the side effects and safety of the drug, what the long term risks and benefits are and to discover how well the drug works when it is used more widely than in clinical trials. More information on clinical trials, it is available at HTTP: www.cancerhelp.org.uk/trials/types-of-trials/phase-1-2-3-and-4-trials (accessed 12 March 2010).
48   A recent gene therapy trial in patients with inherited retinal disease yielded positive results, with the patients having modestly increased vision with no side effects. See A. M. Maguire, F. Simonelli, E. A. Pierce, et al., 'Safety and Efficacy of gene transfer for Leber's Congenital Amaurosis', 358 (2008) New England Journal of Medicine, 2240. In 2009 scientists successfully used a modified virus to deliver a therapeutic gene to halt the development of X-linked adrenoleukodystrophy (ALD), an ultimately and rapidly fatal brain disorder, in two boys. See N. Cartier, S. Haecin-Bey-Abina, C. C. Bartholomae et al., 'Hematopoietic Stem Cell Therapy with a Lentiviral vector in X-Linked Adrenoleukodystrophy', 326 (2009) Science, 818.
49   G. T. Laurie, Genetic Privacy A Challenge to Medico-Legal Norms Cambridge: Cambridge University Press, 2002, 93.

information, this approach is simplistic and often unsatisfactory,[50] largely because information about a person's genes has implications for her family as well as herself. Genetic information is not necessarily unique in this way; for many years family history has been recognized as a valuable predictor of disease. However, as Laurie notes, family history is abstract knowledge flawed by bad or failing memories, and a lack of accurate data and understanding.[51] Genetic data injects an element of certainty by identifying the existence of a specific gene or genetic mutation. Thus, if an individual discovers that she possesses a deleterious gene, the question arises as to whether members of her family should be informed that they may also possess the same potentially harmful genetic makeup. [52]

Conclusion

This chapter provides a brief overview of the elements of embryo testing and acquisition of genetic information, and emerging opportunities for treatment and avoidance which are relevant to the novel genomic claims which form the core of this book. It is hoped that this brief and targeted discussion will place the reader in a position to understand the particular genetic services to which the novel claims considered here pertain, and the ways in which mistakes might be made in the performance of these techniques that might lead to novel legal challenge.

50 Ibid. 93.
51 Ibid. 94.
52 The claim considered in Chapter 7 concerns this issue.

Chapter 2

The recognition of new interests and corresponding duties of care in English negligence law

Introduction

This book analyses the potential reaction of the English courts to claims based on perceived injustices arising from genetic services. In the absence of a dedicated regulatory regime, or a contract, it is argued that these novel genomic grievances would fall to be articulated via the tort system. Here the particular focus is on how English negligence law might react to these grievances.

For the most part, it is argued that the novel claims would seek to establish a new type of damage in the tort of negligence.[1] Damage is the gist of the action in negligence.[2] Historically, tort law is happiest when faced with damage that arises in knotty problems involving collisions between strangers, preferably with lots of broken limbs.[3] The further the perceived adverse outcome is from that corporeal paradigm, the more difficult it becomes to refer to it as an 'injury' in negligence.[4] The perceived harm in the novel genomic negligence claims considered here does not fit within this traditional paradigm of physical harm. The genomic claims would seek to argue that unwanted birth, the birth of the wrong child, the failure to give relevant information or the unsolicited disclosure of unwanted information ought to be recognized as deleterious where they occur as a result of another's negligence.[5] Given that these claims relate to the defeating of interests which are not legally recognized harm in negligence, they are currently largely likely to meet with rejection. Nevertheless, on a theoretical

---

1 The term damage in itself seems to suggest some sort of adverse tangible physical sequelae. The terms damage, injury and loss will be used interchangeably in this chapter to describe the adverse event in the novel genomic negligence claims.

2 J. Stapleton, 'The Gist of Negligence. Part I: Minimum Actionable Damage' 104 (1988), *Law Quarterly Review*, 213.

3 J. Conaghan, 'Tort Law and Feminist Critique' in M. D. A. Freeman (ed.), *Current Legal Problems*, Oxford: Oxford University Press (2004) 174, 192.

4 N. Priaulx, 'Joy to the World! A (Healthy) Child is Born! Reconceptualizing 'Harm' in Wrongful Conception, 13 (2004), *Social and Legal Studies*, 5, rendering an interpretation of J. Feinburg, *Harm to Others,* Oxford: Oxford University Press (1984), 33.

5 See Chapters 4, 5, 6 and 7 respectively.

level, each of the aggrieved parties has suffered a setback to her interests. She has been deprived of ignorance, knowledge, the right child or oblivion through not being born. Alternatively, she might have been somehow affected mentally by the negligent party's actions in a way that falls short of recognized psychiatric harm in the tort of negligence.

This book takes a different perspective regarding the interest that has been interfered with in the genomic negligence claims. It provides a novel focus by examining the implications of recognizing an interest in autonomy as the basis for damage in the tort of negligence. Given the likelihood that these novel claims will be rejected, as negligence law stands today, an explicit recognition of the interest in autonomy might provide a basis for their legal recognition.[6] Autonomy is a fundamental social value; whose import is recognized in some legal contexts. It is well recognized within the tort of trespass, that an intentional failure to respect a patient's autonomous wishes with regard to medical treatment can amount to a battery.[7] The European Convention on Human Rights also provides some protection for autonomy via Article 8.[8] Furthermore, fairly recently, the House of Lords has recognized the need to provide some redress for negligently inflicted setbacks to autonomy interests.[9]

Part I of this chapter focuses on the concept of harm within the tort of negligence. This part analyses the current restrictive culture of English negligence law with respect to the recognition of novel interests which do not fall within the range of recognized harms. The focus is on the concept of duty of care as a mechanism for declining to recognize new interests within the tort of negligence. By focusing on the gate-keeping function of the concept of duty of care, this discussion does not suggest that these novel claims will not present issues with respect to breach, causation and remoteness of damage. However, there is not the space in this book to analyse the novel claims from the perspective of each of the requirements of the tort of negligence; hence the focus on duty.[10] In Part I there is a consideration of how legal harm might be construed with respect to states which might not be universally recognized as deleterious. Part II explains why the interest in autonomy as a basis for the recognition of harm in the genomic claims is not approached from the perspective of the tort of battery.

---

6 In the context of the perceived injustices considered here, the recognition of the interest in autonomy has the potential to provide a comprehensive approach which would require the courts to recognize one new interest as opposed to the recognition of several different types of interest.

7 See for example, *Re B (Consent to Treatment: Capacity)* [2002] 1 FLR 1090; *St George's Healthcare NHS Trust v S* [1998] 3 WLR 936; *R v Collins Ex p. S* (No. 2) [1999] Fam. 26.

8 See for example, *NHS Trust A v M, NHS Trust B v H* [2001] Fam. 348; *R (On the Application of Purdy) v DPP* [2009] UKHL 45, Lord Hope, 34–43.

9 See *Rees v Darlington Memorial Hospital NHS Trust* [2004] 1 AC 309 and *Chester v Afshar* [2005] 1 AC 134.

10 It is assumed that the defendant has acted negligently and her actions are the cause of the set back to the interest.

The concept of harm in the tort of negligence

Given that damage is the gist of the action in negligence,[11] it is crucial to their success that claimants are able to demonstrate interference with an interest which the law recognizes as deleterious. Despite this, many prominent academic commentators claim that the concept of damage in tort is underdeveloped.[12] Stapleton argues that although the issue of what constitutes minimum actionable damage is crucial to an understanding of the limits of the law of negligence, the issue is rarely addressed squarely by the courts.[13] Nolan agrees, arguing that given that damage is an essential component of negligence liability, it is strange that it should be so widely ignored.[14]

Traditionally, the tort of negligence protects physical interests. However, the tort developed out of an action on the case, thus from its beginnings, it did not have inherent boundaries demarcating what might be recognized as deleterious. Nowadays negligence generally follows a pattern of evolution that consists of one step forward and two steps back.[15] As societies become more developed, they become more willing to look sympathetically on novel complaints.[16] Furthermore, developments at the supranational level bring into sharp focus the question of the kind of interests that should be afforded domestic legal protection.

The evolution of duty and damage in English negligence law

The nature of the damage in negligence is crucial to the character and scope of the duty of care owed.[17] The definition and categorization of damage tends to be dealt with under the concept of duty of care. Nolan believes that issues concerning actionable damage are frequently repackaged as questions of

---

11 J. Stapleton, 'The Gist of Negligence. Part I: Minimum Actionable Damage' 104 (1988), *Law Quarterly Review*, 213.

12 J. Stapleton, 'The Gist of Negligence. Part I: Minimum Actionable Damage' 104 (1988), *Law Quarterly Review*, 213; C. von Bar, 'Damage Without Loss' in *The Search for Principle*, W. Swadling and G. Jones (eds), Oxford: Oxford University Press (1999). For support see D. Nolan, 'New Forms of Damage in Negligence' 70 (2007), *Modern Law Review* 59, 59.

13 For a notable exception in the English courts see *Caparo Industries plc v Dickman* [1990] 2 AC 605, Lord Bridge, 618, 627; Lord Oliver, 651.

14 D. Nolan, 'New Forms of Damage in Negligence' 70 (2007) *Modern Law Review* 59, 59–60. Nolan argues that there has been both judicial and academic neglect of the concept of actionable damage.

15 S. Deakin, A. Johnston and B. Markesinis, *Tort Law*, fifth edition, Oxford: Oxford University Press (2003), 2.

16 Ibid. 1.

17 For explicit judicial pronouncements of this fact see, for example, *Sutherland Shire Council v Heyman* (1985) 59 ALR, Brennan J, 564, 590; *Caparo Industries plc v Dickman* [1990] 2 AC 605, Lord Bridge 618, 627: 'It is always necessary to determine the scope of the duty by reference to the kind of damage' and Lord Oliver, 651.

duty.[18] Witting also argues that the existence of a duty of care depends on
the kind of damage pleaded.[19] He argues that little more than foreseeability
must be proven to establish that a duty of care is owed in cases of physical
injury to a person, or physical damage to property, and primary victims who
suffer recognized psychiatric injury.[20] However, more onerous duty requirements apply when the defendant is alleged to have caused secondary victim
psychiatric injury, or economic loss.[21] Where the courts believe there is an
absence of recognizable damage, they explain their decision not to recognize
the novel interest by presenting the situation as one where the defendant
does not owe a duty of care.[22] This is not to say that other elements of negligence such as causation and breach of duty cannot be used as a mechanism
for rejecting claims which do not plead recognizable damage. However, the
concept of duty of care is the most effective at controlling the boundaries of
liability in negligence whether on the basis of limiting the range of recognizable injury or otherwise.[23] Aided by a generational change in the House
of Lords, a more restrictive approach to duty of care emerged in the early

18 D, Nolan, 'New Forms of Damage in Negligence' 70 (2007), *Modern Law Review* 59, 59.

19 C. Witting, 'Physical Damage in Negligence' 61 (2002), *Cambridge Law Journal* 189, 189.

20 Ibid.

21 Ibid.

22 See, for example, *McKay v Essex AHA* [1982] QB 1166 and *McFarlane v Tayside* [2000] 2 AC 59 for
two cases where the courts refused to recognize damage by refusing to recognize that the defendant
owed a duty of care, which are particularly relevant to the discussion in this book. See also *Cowan v
Chief Constable of Avon and Somerset Constabulary* [2001] EWCA Civ 1699 where the Court of Appeal
held that the police owed no duty of care to a tenant unlawfully evicted in their presence but did not
mention the fact that the claimant did not appear to have suffered any actionable damage and
Elguzouli-Daf v Commissioner of Police of the Metropolis [1995] QB 335.

23 The concepts of breach and causation have not proven themselves as quite so effective in controlling
the boundaries of liability. In recent years, the House of Lords has relaxed the causal requirements
where, as Lord Nicholls explained; 'this is what justice requires and fairness demands'. *Fairchild v
Glenhaven Funeral Services Ltd* [2003] 1 AC 32, Lord Nicholls, 68. This relaxation has led to a significant extension of the scope of potential liability in negligence. See, in particular, *Fairchild v Glenhaven Funeral Services Ltd* [2003] 1 AC 32 and *Chester v Afshar* [2005] 1 AC 134 both of which
remove the need for the claimant to prove a direct link between the defendant's negligence and the
claimant's damage in order to establish liability. Issues such as the identity of the defendant seem to
play a part in their Lordships' willingness to relax the causal hurdles, which has resulted in a rather
patchy application of the relaxed requirements. See, for example, *Wilsher v Essex Area Health Authority*
[1988] AC 1074; *Hotson v East Berkshire HA* [1987] AC 750 and *Gregg v Scott* [2005] 2 AC 176. In
the context of claims against local authorities, the prospect of shifting emphasis from issues of duty
to issues of breach in dividing claims which ought reasonably to lead to recovery and claims which
ought not was advocated by Lord Bingham in *JD v East Berkshire Community Health NHS Trust; MAK
v Dewsbury Healthcare NHS Trust; RK v Oldham NHS Trust* [2005] 2 AC 373, 400. His Lordship felt
such a shift would be welcome because 'the concept of duty had proved itself as a somewhat blunt
instrument for facilitating such a division'. However, the remainder of the House did not share Lord
Bingham's enthusiasm for such a shift in emphasis. Lord Nicholls felt that an 'abandonment of the
concept of duty of care in English law, unless replaced by a control mechanism which recognizes this
limitation, is unlikely to clarify the law', 409.

1990s,[24] providing a mechanism for fixing the boundaries of the tort of negligence.[25] However, the court's desire to use the principle to restrict the boundaries of liability in negligence has fluctuated, generating periods of expansion and restriction.

The starting point for the existence of the general duty of care in negligence is *Donoghue v Stevenson*. Lord Atkin laid down the principle that if the harm was reasonably foreseeable and there was a degree of proximity between the parties, in terms of a close and direct relationship,[26] a duty of care could be owed. However, little was said about how the principle might be confined; it was not limited by reference to particular types of loss. Over the years, the courts have sought to refine Lord Atkin's principle to contain the law within the boundaries of manageability. Following *Donoghue*, the courts held that the principle only applied to physical damage to person or property caused by a manufacturer's negligence.[27]

However, from the 1960s to the 1980s, a broader approach to the types of loss that might be recognized in negligence arose, bringing various situations which had little in common with the snail in the ginger beer scenario within the remit of the tort of negligence. The first radical extension of liability came in 1964 in *Hedley Byrne v Heller and Partners*,[28] when the House of Lords recognized for the first time that a duty of care might be owed in negligence for financial loss caused by a careless misstatement. Two further House of Lords' decisions in the 1970s set the law on an expansive path. *Home Office v Dorset Yacht Co Ltd*[29] established that a duty of care might be imposed on an individual to prevent damage being caused by third parties. Thus, liability need not be based on a positive act but could be imposed on the basis of a mere omission. According to Lord Reid, the time had come when 'Lord Atkin's neighbour principle ought to apply, unless there is some justification or valid explanation for its exclusion'.[30] This expansive approach was confirmed in *Anns v Merton LBC*,[31] Lord Wilberforce said:

24 See, for example, D. Howarth, 'Negligence After *Murphy*: Time to Rethink' 50 (1991), *Cambridge Law Journal* 58, 59. See also B. S. Markesinis and S. Deakin, 'The Random Element of their Lordships' Infallible Judgment: An Economic and Comparative Analysis of the Tort of Negligence from *Anns* to *Murphy*' 55 (1992), *The Modern Law Review* 619, 620.

25 In *Home Office v Dorset Yacht Co.* [1969] 2 QB 412, 426, Lord Denning: 'This talk of 'duty' or 'no duty' is simply a way of limiting the range of liability for negligence'; B. S. Markesinis and S. Deakin, 'The Random Element of their Lordships' Infallible Judgment: An Economic and Comparative Analysis of the Tort of Negligence from *Anns* to *Murphy*' 55 (1992), *Modern Law Review* 619, 642.

26 See *Donoghue v Stevenson* [1932] AC 562, Lord Atkin, 580.

27 See, for example, *Farr v Butters Bros. & Co.* [1932] 2 KB 606, Scrutton LJ, 613; *Deyong v Shenburn* [1946] KB 227, Du Parcq LJ, 233.

28 *Hedley Byrne v Heller and Partners* [1964] AC 465.

29 *Home Office v Dorset Yacht Co Ltd* [1970] AC 1004.

30 Ibid. Lord Reid, 1027.

31 *Anns v Merton LBC* [1978] AC 728.

The position has now been reached that in order to establish that a duty of care arises in a particular situation, it is not necessary to bring the facts of that situation within those of previous situations in which a duty of care has been held to exist.[32]

His Lordship suggested that in determining whether a duty of care exists, the courts should ask whether there is, between the claimant and defendant, a sufficient relationship of proximity or neighbourhood, such that the defendant can 'reasonably foresee that her carelessness may be likely to cause damage to the claimant'.[33] If the answer is yes, then it becomes necessary to consider whether there are 'any considerations which ought to negative or reduce or limit the scope of the duty or the class of person to whom it is owed, or the damages to which a breach of it may give rise'.[34] This signalled the rejection of an approach whereby the claimant had to demonstrate analogy with an established category of duty, thereby enabling recognition of the defeating of interests which had not been previously recognized within the tort of negligence; in particular, harm by way of psychiatric damage and pure economic loss caused by a negligent act.[35]

However, a retreat from the expansive approach was initiated by Lord Keith in the mid 1980s in *Governors of the Peabody Donation Fund v Sir Lindsay Parkinson Co. Ltd.*[36] His Lordship suggested that a tendency to treat the two stage formula in *Anns* as definitive had emerged, which should be resisted.[37] Following this, significant hostility was demonstrated towards the *Anns* test.[38] *Anns* was finally overruled in *Murphy v Brentwood DC*.[39] The House noted that The High Court of Australia declined to follow *Anns* in *Council of the Shire of Sutherland v Heyman*[40] on the basis that:

---

32 Ibid. Lord Wilberforce, 751–752.

33 Ibid.

34 Ibid.

35 See, for example, the expansive approach taken to nervous shock in *McLoughlin v O'Brien* [1983] AC 410 and economic loss in *Junior Books v Veitchi* [1983] AC 520.

36 *Governor of the Peabody Donation Fund v Sir Lindsay Parkinson Co. Ltd* [1985] AC 210.

37 Ibid. Lord Keith, 240.

38 See, in particular, *Leigh and Sillavan Ltd. v Aliakmon Shipping Co. Ltd.* [1986] AC 785, Lord Brandon, 815. *Curran v Northern Ireland Co-ownership Housing Association Ltd.* [1987] AC 718, Lord Bridge, 724–725. *Yuen Kun Yeu v Att. Gen. of Hong Kong* [1988] AC 175, Lord Keith, 194. *Hill v Chief Constable of West Yorkshire* [1989] AC 53 and *D & F Estates Ltd v Church Commissioners for England* [1989] AC 177. See also, J. C. Smith and P. Burns, '*Donoghue v Stevenson* – The Not So Golden Anniversary' 46 (1983), *Modern Law Review* 47, where the authors criticize the trend in the law of negligence towards the elevation of the 'neighbourhood principle' into one of general application from which a duty of care may always be derived unless there are clear countervailing considerations to exclude it.

39 *Murphy v Brentwood DC* [1991] 1 AC 398.

40 *Council of the Shire of Sutherland v Heyman* (1985) 157 CLR 424.

It is preferable … that the law should develop novel categories of negligence incrementally and by analogy with established categories, rather than by a massive extension of a prima facie duty of care restrained only by indefinable considerations which ought to negative, or to reduce or limit the scope of the duty or the class of person to whom it is owed.[41]

Lord Keith opined that, in novel duty situations, an incremental approach along these lines was to be preferred to the two-stage *Anns* test.[42] Thus began a new era in the tort of negligence, whereby novel claims had to be brought within an existing analogous category of liability before a duty of care might be owed.

Caparo v Dickman[43] was decided shortly after *Murphy*. Here the House held that in order to establish a duty of care, the claimant must demonstrate that her case was covered by a direct, or closely analogous, precedent where a duty of care had already been imposed. If there is no such authority, the court could apply three criteria to determine whether a duty of care exists.[44] Under this approach, significant emphasis is placed on the concept of reasoning by analogy, thereby curtailing the development of the law by limiting liability to scenarios where liability has previously been recognized. The restrictive *Caparo* approach to the question of duty of care prevails to this day. Nevertheless, this does not mean that the law is incapable of recognizing new interests in negligence. In *Donoghue v Stevenson*, Lord Macmillan was clear that 'the categories of negligence are never closed'.[45] According to Oliphant, tort law can be employed to protect whatever interests are deemed worthy of protection in any particular society: the list of protected interests is not set in stone.[46] Nevertheless, it is likely to be more difficult for a claimant to establish a new duty of care with respect to a novel interest under *Caparo* than it would have been under *Anns*.

Some torts are actionable per se; that is without proof of damage. Trespass and libel are the most prominent forms of such action. It has been suggested that the fact that some torts are actionable per se demonstrates that concern for basic human rights has always been a concern of English tort law.[47] This raises the question of whether tort law is a method of repairing harm or vindicating rights. Traditionally, negligence in particular has been seen as

---

41 Ibid. Brennan J, 481.
42 *Murphy v Brentwood DC* [1991] 1 AC 398, Lord Keith, 461.
43 *Caparo Industries plc v Dickman* [1990] 2 AC 605.
44 Generally known as the three stage test of what is foreseeable, proximate and just, fair and reasonable.
45 *Donoghue v Stevenson* [1932] AC 562, Lord Macmillan, 619.
46 K. Oliphant, 'The Nature of Tortious Liability' in A. Grubb (ed.) *The Law of Tort*, London: Butterworths (2002), para 1.12.
47 C. von Bar, 'Damage without Loss' in W. Swadling and G. Jones (eds) *The Search for Principle*, Oxford: Oxford University Press (1999), 30.

doing the former. However, more recently the courts seem to be using the language of rights to justify recognition of novel claims in negligence.[48] Weir argues:

> By using the concept of 'right' we can gloss over the fact that no damage need be proved. It might however, be better to admit that in addition to its more obvious function of redressing harms the law of tort also vindicates rights: it has a constitutional as well as a compensatory function.[49]

As rights become a more important and prominent feature of modern life, the distinction between redress for harm and vindication of rights becomes more difficult to discern. If rights are so important, it might be argued that breach of a right constitutes a harm. As Christian von Bar argues, the violation of a right such as the right to vote can constitute damage.[50] Other commentators agree that the violation of rights cannot be distinguished from damage.[51] This book focuses on the tort of negligence, where it has long been recognized that damage is the gist of the action. Although damage is not a concept which only denotes tangible physical harm,[52] the tort of negligence is not a system which recognizes the pure vindication of interests.[53] Currently, where interests are negligently violated, further loss must be demonstrated before an action in negligence will lie. However, what constitutes a loss remains open to question.[54] It might be argued that the loss is inherent in the interference with a person's fundamental rights. Furthermore, if the courts impose a duty of care with regard to a particular right, it might be argued that the duty is emptied of all content if it is not inherent in the breach of that duty that a harm exists.[55] Indeed, Stoll argues that the only workable definition of damage in the legal sense is a detriment that the

---

48 *Rees v Darlington Memorial Hospital NHS Trust* [2004] 1 AC 309 and *Chester v Afshar* [2005] 1 AC 134.

49 T. Weir, *A Casebook on Tort*, seventh edition, London: Sweet and Maxwell (1994), 428–429.

50 C. von Bar, 'Damage without Loss' in W. Swadling and G. Jones (eds) *The Search for Principle*, Oxford: Oxford University Press (1999), 28.

51 De Cupis, *Il Danno Teoria generale della Responsabilitá Civile* ii third edition, Milan, 1979, 232 cited in C. von Bar, 'Damage without Loss' in W. Swadling and G. Jones (eds) *The Search for Principle*, Oxford: Oxford University Press (1999), 27.

52 The tort of negligence has for a long time now recognized the interests in mental and financial well-being and, as this chapter demonstrates, there is evidence that the tort of negligence is capable of recognizing other intangible sequelae as damage.

53 The discussion in this book speaks of autonomy as an interest rather than a right on the basis that this book does not provide a rights-based analysis of the potential legal response to interference with autonomy.

54 See the discussion below.

55 That is if the claimant has to demonstrate further adverse consequences that fit within the traditional paradigm of damages within negligence. See, for example *Chester v Afshar* [2005] 1 AC 134, Lord Hope, 162–163; Lord Walker, 166.

legal system regards as damage to be compensated in a specific form according to tort provisions.[56] From this perspective, damage is the event, the prevention of which is the purpose of the duty. Or, as von Bar argues, 'damage is the reverse of duty'.[57] Thus, an analysis of what amounts to legal damage finds its focus in the concept of duty of care.

An expansive approach to the imposition of duty of care with respect to novel interests?

The restrictive approach in *Murphy* and *Caparo* continues to influence the judiciary to be cautious about recognizing setbacks to novel interests. In *X (minors) v Bedfordshire CC* the House of Lords considered five appeals, three of which made allegations of defective educational provision on behalf of local educational authorities with respect to children with special educational needs. The case was struck out as disclosing no reasonable cause of action, leading to the conclusion that defective education was not legally recognizable harm. However, fuelled by the Human Rights Act 1998 and the jurisprudence of the European Court of Human Rights (ECtHR), a more expansive approach emerged with regard to the kind of interests that could be recognized in negligence.

The initial restrictive approach to the recognition of an interest in avoiding impaired educational development which was taken in *X (minors)* was, following three ECtHR decisions,[58] challenged in *Phelps v Hillingdon LBC*.[59] All four of the claimants in *Phelps* alleged that there was a negligent failure to provide them with adequate educational provision for their special needs. The House unanimously allowed the appeals. Lord Slynn said that failure to diagnose a congenital injury and to take appropriate action as a result of which a child's level of achievement is reduced, may amount to 'damage for the purpose of the common law'.[60] In effect, the decision recognizes that the

---

56 H. Stoll, *Haftungsfolgen im Bürgerlichen Recht*, Heidelberg, (1993), 207, 239 cited in C. von Bar, 'Damage without Loss' in W. Swadling and G. Jones (eds) *The Search for Principle*, Oxford: Oxford University Press (1999), 29.

57 C. von Bar, 'Damage without Loss' in *The Search for Principle* W. Swadling and G. Jones (eds), Oxford: Oxford University Press (1999), 29.

58 *TP and KM v United Kingdom* (2001) 34 EHRR 42 and *Osman v UK* (1999) 29 EHRR 245. The ECtHR subsequently conceded in *Z v UK* that the approach in *Osman* had been wrong, but the *Osman* ruling already appeared to have had an effect on decisions in the domestic courts. See, for example, *Barrett v Enfield LBC* [2001] 2 AC 550, Lord Browne-Wilkinson, 557–560; *S v Gloucestershire CC* [2001] Fam. 313, May LJ, 339–340; *L (a child) and Another v Reading Borough Council and Another* [2001] 1 WLR 1575, Otton LJ, 1585–1588. The courts have indicated relatively recently that the 'uncertain shadow of *Osman* still lies over this area of the law' in *Matthews v Ministry of Defence* [2003] 1 AC 1163, Lord Walker, 1127.

59 *Phelps v Hillingdon LBC* [2001] 2 AC 619.

60 Ibid. Lord Slynn, 654.

continuing effects of a failure to ameliorate a condition such as dyslexia, are themselves a form of actionable damage.[61]

However, the expansive approach to the question of whether negligent defective educational provision amounts to a recognizable interest has not been consistent. In the Court of Appeal in *Phelps*, Stuart-Smith LJ said that dyslexia itself was not an injury and he could not therefore see how failure to ameliorate its effects might be an injury either.[62] Furthermore, although the central issue in *Phelps* was whether the setback to the interest in not reaching one's potential because of defective educational provision in the context of special educational needs amounted to a recognizable interest, the House in fact glossed over this point.[63] Subsequently, it has become clear that the House's recognition of the interest in impaired educational development is narrow. In *Adams v Bracknell Forest DC*[64] the House of Lords confined *Phelps* to cases where there is some form of disability requiring special educational provision. In *Adams*, Lord Scott was of the opinion that the 'deprivation of the benefit of literacy' does not fit within the concept of a personal injury because it is not an 'impairment of a physical or mental state'.[65] Thus, his Lordship was not prepared to recognize a novel interest based on defective educational development. Nolan argues that following *Adams*, impaired educational development is not *generally* a form of personal injury.[66] Thus, *Phelps* represents an exceptional expansive approach. The indications are that this expansive approach does not apply in other types of cases against public authorities, even where the interest interfered with is one which already receives general legal recognition.[67]

Individual perceptions of harm

Given that the current approach to the recognition of new interests in the tort of negligence is relatively restrictive, we can assume that the novel genomic claims discussed in this book are likely to meet with judicial

---

61 See D. Nolan, 'New Forms of Damage in Negligence' 70 (2007), *Modern Law Review* 59, 81.

62 *Phelps v Hillingdon LBC* [1999] 1 ALL ER 421, Stuart-Smith LJ, 433.

63 D. Nolan, 'New Forms of Damage in Negligence' 70 (2007), *Modern Law Review* 59, 81. The issue was presented as one of whether a duty of care was owed by the defendant educational authority. See above for an in depth discussion of the concept of duty as the mechanism for explaining decisions not to recognize a setback to a particular interest as damage.

64 *Adams v Bracknell Forest DC* [2004] UKHL 29.

65 Ibid. Lord Scott, 68

66 D. Nolan, 'New Forms of Damage in Negligence' 70 (2007), *Modern Law Review* 59, 84 (original emphasis).

67 The House of Lords took a restrictive approach in *JD v East Berkshire Community Health NHS Trust; MAK v Dewsbury Healthcare NHS Trust; RK v Oldham NHS Trust* [2005] 2 AC 373 and dismissed appeals by parents who had been wrongly accused of child abuse by the local authority, confirming that they were not entitled to claim damages for a recognized head of damage; psychiatric harm.

resistance.[68] One of the problems with the novel genomic grievances considered here is that the essence of the grievance rests on the individual's perception of her circumstances, rather than a universal perception of what is harmful. This book considers the claims that might arise when an individual is aggrieved because of her birth, or with respect to the child she has given birth to, or because she was or was not given information about herself.[69] In these circumstances, the situation which the individual complains of is not uniformly experienced as deleterious. Moreover, the problem with defining some of the situations which form the basis of the novel claims considered here as harmful, is that many people would view the situation as positive.[70] Conceptually, it might be difficult to define as harm that which many people would welcome.

The question of whether there is a loss in the genomic claims depends on the perspective of the individual.[71] This is different to the infliction of personal injury or disease or financial loss, which is likely to be uniformly experienced as deleterious. From a practical perspective, it might be argued that it is unfair to hold that a person owes a duty of care with respect to an outcome which is not generally considered to be harmful. However, the tendency to view the kinds of situations described here as not harmful, arises from the historical inability to exercise control over the situation. Before testing for genetic disorders was possible, parents had little ability to avoid the birth of an affected child,[72] or to know about their genetic future. Modern genetic technology provides the means to

---

68 The setbacks in the novel grievances examined here are in the interests of not being born, in having the child one desires, in having relevant information and in remaining in ignorance. None of these interests can be easily assimilated into the range of interests for which the tort of negligence currently provides protection. Although it might be possible to argue that each of these interests could form the basis of an action in negligence, this book seeks to establish a more comprehensive approach to the question of the interest interfered with in novel genomic negligence claims, such as these arising from new genetic technologies. It seeks to argue that the interest to which there has been a setback in all of these cases is that of autonomy. Thus, if the courts recognized the interest in autonomy, it would provide a basis for the recognition of all of the claims considered here and many others which might arise from carelessness in the handling of genetic material or genetic information.

69 The argument in this book is that all of these perceived grievances could be interpreted as interferences with the individual's autonomy. This issue is considered in detail in Chapter 3 and each of the chapters exploring the novel claims.

70 For example, the birth of a child is usually a happy occurrence. However, this is not always the case. Many people who become pregnant might not want to be pregnant, but with the wide availability of contraception and abortion, the majority of people who do not desire to give birth to a child will not do so.

71 In *Harriton v Stephens* [2006] HCA 15, 168, Hayne J thought that the question of whether there is damage directs attention to the position of the particular plaintiff, not some hypothetical class of persons of which the plaintiff might be said to be a member.

72 Other than not reproducing or relying on the then little understood implications of family history, people may have had little indication of their own deleterious genetic traits and the implications of those traits for their offspring.

prevent the conception and birth of individuals with genetic disorders and enables people to know intimate details about their own genome, providing valuable information about their future. However, people's views regarding testing offspring for genetic disorders or knowing or not knowing about genetic risks that they can or cannot do something about, will be personal and varied.

Nevertheless, this does not lead to the conclusion that outcomes of genetic technology which are not universally perceived as detrimental should be incapable of being perceived as harm. In analysing the potential of pregnancy[73] to constitute a 'harm', Nolan argues that pregnancy is a state which is usually perceived as beneficial but also as capable of amounting to a harm.[74] Mullis agrees that pregnancy may be a personal injury in some cases but not in others.[75] Even though some women are pleased when they find out that they are pregnant, this does not mean that others do not perceive pregnancy as harmful. One of the problems associated with notions of damage which have this subjective element is that the question of damage depends not only on what takes place, but also on the claimant's attitude to it.[76] Thus, there is a possibility that a defendant could be held liable for causing an outcome which the claimant considers harmful, but which the defendant could not reasonably have foreseen would be perceived in such terms.[77] However, where the harm is based on the claimant's perception of the event as such, there may be actual or imputed knowledge of what the individual desired prior to the negligent act which makes the act a wrong and leads to the view that its consequences, although not universally perceived as deleterious, were reasonably foreseeable as deleterious to the particular individual.[78]

Even if the defendant did not know what the individual claimant's attitude to an outcome would be before the carelessness occurred, it does not rule out the possibility that causing the outcome which is not always recognized as harmful should, therefore, never be perceived to be harmful. This is particularly the case if there is regulation permitting and legitimizing the outcome, which the negligence frustrates, which might indicate its deleterious nature to the particular individual. Even though some outcomes might not traditionally be recognized as harmful because of a lack of ability to control the outcome, the ability to control that outcome might influence people to recognize that in certain circumstances the outcome is a harm.

---

73 A state not universally perceived as detrimental but, nonetheless, unwanted in many instances.

74 D. Nolan, 'New Forms of Damage in Negligence' 70 (2007), *Modern Law Review* 59, 71–77.

75 A. Mullis, 'Wrongful Conception Unravelled' 1 (1993), *Medical Law Review* 320, 325.

76 D. Nolan, 'New Forms of Damage in Negligence' 70 (2007), *Modern Law Review* 59, 71–77.

77 A. Grubb, 'Failed Sterilisation: Limitation and Personal Injury' 4 (1996), *Medical Law Review* 94, 97.

78 The claimant may have approached the defendant to provide her with legitimate assistance in achieving the particular outcome which is, due to the defendant's carelessness, not achieved.

Societies are progressive and have a tendency to become more sympathetic to a wider spectrum of loss as they develop. [79] The more able we are to produce more and greater harm; it seems the less willing we are to put up with losses, and more willing to assert our rights in a legal way.[80] An increase in the number of claims presenting a particular outcome as harm is perhaps an indication that society is beginning to recognize that the loss in question ought to be of high enough priority for legal recognition.[81] It is possible, over the years, to trace a change in judicial attitudes to the kinds of harm that might be of sufficiently high enough priority for compensation, perhaps influenced by changes in societal attitudes as to what constitutes harm.

Psychiatric injury was not a traditionally recognized harm in negligence.[82] However, as doctors and society began to accord higher priority to mental security, the law followed suit and the scope of the duty of care in negligence expanded to recognize that psychiatric injury could amount to a legally recognized loss.[83] Nevertheless, the courts continue to use the principle of duty of care to draw a distinction between physical and psychiatric harm, only allowing recovery for the latter in special circumstances. Weir argues that the majority of people would draw a distinction between physical and psychiatric damage in terms of the provision of legal remedies:

> There is ... no doubt that the public draws a distinction between the neurotic and the cripple, between the man who loses his concentration and the man who loses his leg. It is widely felt that being frightened is less than being struck, that trauma to the mind is less than lesion to the body. Many people would consequently say that the duty to avoid injuring strangers is greater than the duty not to upset them. The law has reflected this situation as one would expect, not only by refusing damages for grief altogether, but by granting recovery for other psychical harm only late and grudgingly, and then only in very clear cases.[84]

In the eighteen years since this observation was made, advances in scientific understanding and shifts in social perceptions of damage have made the division between physical and psychiatric harm increasingly difficult to draw

---

79 S. Deakin, A. Johnston and B. Markesinis, *Tort Law*, fifth edition, Oxford: Oxford University Press (2003), 1.

80 Ibid. 2–3.

81 For the courts' perception of prevailing social views as relevant to the question of the recognition of new types of harm in negligence see, for example, *Turpin v Sortini* (1982) 643 P 2d 954, Kaus J, 961; *McFarlane v Tayside Health Board* [2000] 2 AC 59, Lord Steyn, 82; Lord Hope, 96; Lord Millett, 114; *White v Chief Constable of South Yorkshire Police* [1999] 2 AC 455, Lord Hoffmann, 510.

82 Unless the claimant had also suffered some physical injury: *Victorian Railway Commissioners v Coultas* (1888) 13 App. Cas. 222.

83 V. Harpwood, *Modern Tort Law*, sixth edition, Cavendish Publishing Ltd (2005), 37.

84 T. Weir, *A Casebook on Tort*, seventh edition, London: Sweet and Maxwell (1992), 88.

with conviction.[85] Lord Steyn has even suggested that the latter may well be far more debilitating than the former.[86] Significant steps have been taken with regard to the recognition that interference with a person's mental well being occasions harm to that person.[87] However, the law has, thus far, declined to put recovery for psychiatric harm on the same footing as physical injury. Nevertheless, an outcome which was once not a harm, can come to be perceived as such. If this is the case, the question of when and whether it is a harm might depend on individual perception. In the absence of the ability to determine individual perception pre-tort, an action based on the new form of harm may seem to the defendant to be based on the whim of the claimant. However, where the defendant's actions are, because of regulation or established norms, deemed to be culpably careless and the consequences of those actions cause an outcome to another which she perceives to be harmful, the law should consider whether she has, in fact, been so harmed.

The interest in autonomy as a basis for liability from the perspective of the tort of battery

This chapter focuses on the fact that the novel genomic grievances considered here concern outcomes which do not currently amount to legally recognized damage. As damage is the gist of the action in negligence, recognition of the novel claims would require an extension of the recognition of what amounts to damage. It is argued in this book that the claimants in each of the novel scenarios have suffered a setback to their interest in autonomy. Here it is argued that English negligence law could be imbued with a specific recognition of the interest in autonomy as a means of recognizing the novel genomic claims. A further possible argument is that an extension of trespass to the person could be made to cover negligent conduct.[88] The

---

85   S. Deakin, A. Johnston and B. Markesinis, *Tort Law*, fifth edition, Oxford: Oxford University Press (2003), 96.

86   *White v Chief Constable of South Yorkshire Police* [1999] 2 AC 455, Lord Steyn, 492.

87   See, for example, *North Glamorgan NHS Trust v Walters* [2003] PIQR P16; *Froggatt v Chesterfield* [2002] All ER (D) 218; *Farrell v Avon Health Authority* [2001] All ER (D) 17; *AB and Others v Leeds Teaching Hospital* [2004] EWHC 644. However, a distinction continues to be drawn between medically recognized psychiatric harm and emotional distress. See *Johnston v NEI International Combustion Ltd; Rothwell v Chemical & Insulating Co Ltd; Topping v Benchtown Ltd; Grieves v F T Everard & Sons* [2007] UKHL 39.

88   In *Fowler v Lanning* [1959] 1 QB 426, 433, Lord Diplock left open the question of whether the courts would recognize a tort of negligent trespass to the person. However, Lord Denning refused to recognize that there was such a thing as unintentional trespass in *Letang v Cooper* [1965] 1 QB 232, 240, where he said 'when the injury is not inflicted intentionally but negligently, I would say the only cause of action is negligence not trespass'. This seems to suggest that there should be an intention to cause the consequences of the unwarranted interference. However, more recent authority demonstrates that whilst the interference must be intentional, the defendant need not intend the consequences of the interference. See, for example, *Williams v Humphrey, The Times* (February 20, 1975).

only trespass tort which offers a potentially viable alternative to the tort of negligence in relation to genomic negligence claims is battery, thus this discussion focuses exclusively on the tort of battery.

Battery is the direct and intentional application of force to another person without that person's consent.[89] The significant benefit in framing the actions considered here in battery is that battery is actionable per se. Rather than providing reparation for consequences, the tort of battery seeks to vindicate the individual's important right to be free from unwanted physical interference. On the face of it, this focus on rights vindication might assist the claimants in the genomic negligence actions who will currently struggle to found their claims in negligence because of the centrality of the concept of damage. However, even though battery is actionable per se, there are many other drawbacks to bringing the novel claims seeking to establish liability for interference with the interest in autonomy in the tort of battery.[90]

First, irrespective of the independent merits of analysing the novel actions in negligence or battery given the elements of each of the tort, there is an advantage to examining the potential outcomes of the claims from a negligence perspective because variants of the genomic claims considered in Chapters 4 and 5 have previously been pleaded in negligence in this, and other, jurisdictions.[91] This provides some evidence of existing negligence law's potential reaction to these types of claims, providing the building blocks for the negligence-focused discussion of the genomic variants of these claims. However, there are more fundamental grounds upon which it might be argued that it is preferable to analyse the novel claims and the argument that they could be legally recognized as interferences with the interest in autonomy, from the perspective of negligence as opposed to battery.

Consent operates as a defence to the tort of battery. This defence has been particularly important within the field of medical treatment. Patients are asked to sign consent forms before they undergo invasive treatment. Consent to less invasive interventions might be presumed in the face of the conduct of the patient.[92] It is on the basis of consent that doctors do not

---

89   *Fagan v Metropolitan Police Commissioner* [1969] 1 QB 439.

90   In this book I have chosen not to analyse the potential for the interest in autonomy to form the basis for the novel action from the perspective of an action in battery. This is not to say that the novel genomic claim considered in Chapter 7 should not be pleaded as a novel battery action. However, the other grievances simply do not involve battery.

91   See the wrongful life type claim considered in Chapter 4 and the wrongful conception/birth type claim considered in Chapter 5.

92   Consent forms are not generally relied upon with respect to minimally invasive and minimally risky interventions such as vaccinations and recording vital signs such as blood pressure, temperature and pulse. The patient's oral consent and physical manifestations are taken to indicate her consent to the intervention.

commit a battery when treating patients. Because consent is a defence to battery, in the medical context the majority of battery actions concern situations where the claimant has refused to give consent, but the doctor treats the patient in any event on the basis that the treatment is in her best interests.[93] Where the claimant does give her consent to a medical intervention, that consent must be real. Consent can be vitiated by fraud or misrepresentation, but a failure to give the relevant information about a procedure or committing some mistake during a procedure does not establish a lack of consent, thereby grounding an action in battery. This distinction goes to the heart of the question of interferences with consent that might amount to battery and interferences with consent that might amount to negligence. As Markesinis and Deakin note, a doctor who fails to give a patient full information prior to an operation will not necessarily be liable in trespass because the question of whether the defendant conformed to the necessary standard of care in advising the patient is separate from the question of whether the patient has given her consent to surgery.[94] In *Chatterton v Gerson*, Bristow J held that more was needed to vitiate consent than a failure of communication between the doctor and patient, which might amount to breach of duty in negligence.[95] According to Bristow J:

> once the patient is informed in broad terms of the nature of the procedure which is intended, and gives her consent, that consent is real, and the cause of action on which to base a claim for failure to go into the risks and implications of treatment is negligence, not trespass.[96]

Markesinis and Deakin argue that the point is that the patient's consent to being operated on is broadly effective to protect the surgeon in respect of that type of operation.[97] The surgeon who performs a circumcision after obtaining consent for a tonsillectomy would probably not be protected by the consent to the tonsillectomy.[98] However, where the surgeon makes a

---

93 Often on a mistaken assumption that the claimant's unwise decision somehow demonstrates her incapacity. However, the difference between incapacity and making unwise decisions is now clearly demonstrated in the Mental Capacity Act 2005, s. 1 (4). A person is not to be treated as unable to make a decision merely because he makes an unwise decision. For an example of such cases see *Re B (Consent to Treatment: Capacity)* [2002] 1 FLR 1090; *St George's Healthcare NHS Trust v S, R v Collins Ex p. S (No. 2)* [1999] Fam. 26.

94 S. Deakin, A. Johnston and B. Markesinis, *Tort Law*, fifth edition, Oxford: Oxford University Press (2003), 418.

95 *Chatterton v Gerson* [1981] QB 432, Bristow J, 442.

96 Ibid. Bristow J, 443.

97 S. Deakin, A. Johnston and B. Markesinis, *Tort Law*, fifth edition, Oxford: Oxford University Press (2003), 418.

98 *Chatterton v Gerson* [1981] QB 432, Bristow J, 443.

careless error leading to the failure of the operation for which there was consent, the claimant's complaint is not that he was operated on against his will, but that the outcome of the operation was detrimental to him.[99]

Thus, where the claimant consents to an operation and she suffers a setback to her interests because the operation is negligently performed, any action brought on her behalf, especially if the setback to that interest constitutes already recognized legal damage, lies in negligence rather than in trespass. The requirement in battery that there be no consent to the interference per se will be problematic for most of the novel claims considered in this book. Chapters 4 and 5 concern novel claims which might arise from a careless failure to properly carry out a genetic service to which the claimant has consented, rather than an interference in the face of a lack of consent. These claims concern carelessness in the procedure to which the claimant consented which does not ground an action in battery, but may ground an action in negligence.

The claim discussed in Chapter 6 is particularly problematic from a battery perspective. This claim concerns a failure to disclose genetic information to a person to whom it is relevant. This is not a grievance which is based on interference without consent. Moreover, this is not a claim based on interference at all; the claimant's grievance effectively concerns a lack of interference. As battery is the direct and intentional application of force to another person without that person's consent, some form of interference is probably crucial. McLean argues that trespass actions cannot cover cases where the failure to have proper regard for the patient's right to make a decision involves the provision of no therapy, or failure to disclose therapeutic alternatives.[100] Although there might be some mileage in discussing whether the tort of battery could be extended to cover psychological interference as well as physical interference,[101] there is little point in considering whether a battery is committed where there is no interference at all. Indeed, this claim concerns a pure omission which is not sufficient to ground liability in trespass to the person.[102]

The claim in Chapter 7 is the only claim that could conceivably be framed as a battery action. This claim is based on an unwanted disclosure

99   S. Deakin, A. Johnston and B. Markesinis, *Tort Law,* fifth edition, Oxford: Oxford University Press (2003), 418.

100   S. McLean, *A Patient's Right to Know: Information Disclosure, the Doctor and the Law,* Aldershot: Dartmouth (1989), 167.

101   Although the historical position in the tort of assault was that words alone could not amount to an assault: *Meads v Belt's case* (1823) 168 ER 1006, in *R v Ireland* [1998] AC 147 (a criminal assault case), the House of Lords denied centuries of precedent and held that assault could be committed by words alone. Thus, under the criminal law at least there appears to be some conception of psychological assault.

102   S. Deakin, A. Johnston and B. Markesinis, *Tort Law,* fifth edition, Oxford: Oxford University Press (2003), 414.

of genetic information. It might be argued that this type of claim could rest on an interference in the face of a refusal to consent, upon which an action in battery could be founded. The claimant would have to overcome the fact that a battery usually involves a physical interference as opposed to a psychological interference.[103] Nevertheless, despite some difficulties, it might be argued that there are good reasons to consider this type of claim in isolation from the perspective of an action in battery. However, the consideration of the genomic claim in Chapter 7 from a battery perspective is beyond the scope of this book which seeks to present a comprehensive consideration of how the English courts might, through the tort of negligence, consistently and comprehensively recognize that the novel genomic negligence claims considered here have caused grievance on the basis that the defendant's negligence interfered with the claimant's autonomy.

A lack of consent, as opposed to an explicit refusal to consent, is not a complete bar to an action in battery. A lack of consent to battery does not necessarily imply imputed consent. In *Collins v Wilcock*[104] and *Re F*,[105] Lord Goff felt that in the context of an action for battery the law should not adopt complicated legal rules based on implied consent, but should simply exclude liability for conduct generally acceptable in everyday life. In *Re F* he said:

> a broader exception has been created to allow for the exigencies of everyday life: jostling in a street or some other crowded place, social contact at parties such like. This exception has been said to be founded on implied consent, since those who go about in public places, or go to parties, may be taken to have impliedly consented to bodily contact of this kind. Today this rationalization can be regarded as artificial.... For this reason I consider it more appropriate to regard such cases as falling within a general exception embracing all physical contact which is generally acceptable in the ordinary conduct of everyday life.[106]

Even if the refusal to recognize battery in these circumstances is not based on imputed consent, the implication of this rule is that you cannot claim

---

103   But see the House of Lord's decision in *R v Ireland* [1998] AC 147. It might be argued that there is scope to argue that this claim could fall within the exception laid down in *Wilkinson v Downton* [1897] 2 QB 57, but given that this very narrow rule requires that the defendant wilfully does an act which is calculated to cause physical harm to the claimant, it is unlikely to cover the scenario considered here.

104   *Collins v Wilcock* [1984] 1 WLR 1172.

105   *Re F* [1990] 2 AC 1.

106   Ibid. Lord Goff, 72–73.

that you did not consent to certain things. So although it might be the case that there is no imputed consent to a particular interference, it is also the case that there can be no refusal to consent to that intervention; thus, the outcome is the same. Chapter 7 considers the issue of whether negligence law ought to recognize, within the context of it being imbued with an interest in personal autonomy, that people can refuse to know relevant information about themselves. From a battery perspective, it might be argued that receiving relevant information about oneself is part and parcel of everyday life and not something which one can refuse. On this basis, the claim considered in Chapter 7 does not lend itself well to analysis within the context of the tort of battery where the absence of consent is fundamental to the success of the action. However, from the general perspective of the principle of autonomy, there does not appear to be a level of interference below which the right to refuse consent does not exist, so that autonomy is not interfered with. The concept of consent based on autonomy is far less developed in negligence than it is in battery. The building blocks for the argument that there are some interferences which you cannot refuse to consent to because they are part of the exigencies of everyday life do not exist in negligence as they do in battery.

The argument that the novel genomic claims should seek to establish a recognized loss, on the basis that the interest in personal autonomy has been interfered with in the tort of negligence, has a further advantage over seeking to articulate the claims as battery actions. Assuming the wrong of carelessness is made out in all of these claims,[107] they would all seek to do the same thing: establish that interference with autonomy amounts to a recognized loss. This allows a comprehensive discussion of the novel claims from the perspective of the recognition of a new kind of damage. On the other hand, as the discussion in this chapter demonstrates, significant ad hoc tinkering would be required if the tort of battery were to be extended to cover these kinds of claim.[108] It might be argued that given that, in recent times, incremental tinkering with the tort system has led to legal rules which appear inconsistent and arbitrary,[109]

---

107   Negligence is often assumed in cases where a hearing with regard to the contentious legal issue is warranted. Indeed, this was the case in some of the cases discussed in this book: *McFarlane v Tayside Health Board* [2000] 2 AC 59; *Rees v Darlington Memorial Hospital NHS Trust* [2004] 1 AC 309; *Parkinson v St James and Seacroft University Hospital NHS Trust* [2002] QB 266; and *Chester v Afshar* [2005] 1 AC 134.

108   Especially as three quarters of the claims here do not fit the battery paradigm at all.

109   The ad hoc development of the law with respect to secondary victims of psychiatric harm has created legal rules which lack clarity and are often difficult to reconcile, leading Stapleton to suggest that this area is both the best historical example of incremental development and yet also the area where the 'silliest rules' now exist and where 'criticism is almost universal'. J. Stapleton, 'In Restraint of Tort' in *Frontiers of Liability* 2, P. Birks (ed.), Oxford: Oxford University Press (1994), 83, 95.

a wholesale approach based on the recognition of a new type of damage would be preferable.[110]

Moreover, despite some disagreement about its importance, the significance of the principle of autonomy in contemporary bioethics, medicine and law is clear.[111] If autonomy is such an important principle, an explicit legal recognition of the principle ought to take a form which reflects the importance of the interest, thereby providing meaningful protection for those whose autonomy has been interfered with. It might be argued that a legal approach to protection of personal autonomy, which classes violation of the interest in autonomy as damage (in negligence), recognizes the importance of the interest in a way which an approach which is based on protecting the interest in autonomy without recognizing that the interference has occasioned loss does not.[112] If the claimant feels she has suffered harm, she might desire legal recognition of the fact that she has been harmed. As the tort of battery is actionable per se, it could recognize the violation of the interest in autonomy without recognizing that any damage has occurred through the violation of that interest. In this case, any redress awarded to the claimant would be nominal and may leave the claimant feeling that the law is incapable of fully recognizing and compensating her for the consequences of the interference with her interests. If interference with the interest in autonomy is recognized as damage in negligence, the claimant's redress would reflect the fact that interference with autonomy amounts to a loss, rather than simply an interference.

Conclusion

Determining the current culture with regard to the recognition of novel interests and corresponding duties of care is an important factor in determining

110 It is assumed here that consistency is an aim of English negligence law. As McLean puts it: the law needs an appropriate framework and workable guidelines in order to function with internal logic and consistency. S. McLean, *Autonomy, Consent and the Law*, London: Routlege Cavendish (2010), 69.

111 The first three principles of the Mental Capacity Act 2005 are designed to maximize autonomy, demonstrating an increasing effort to ensure that choices are autonomous. Some commentators challenge interpretations of autonomy that inadequately attend to issue of time and community. However, those who take up these themes do not argue that autonomy is not important. They reaffirm a modified version of the role of autonomy in biomedical ethics. For a further discussion of this issue see J. F. Childress and J. C. Fletcher, 'Respect for Autonomy' 24 (1994), *Hastings Center Report*, 34. Some question the primacy of autonomy in medical law. See, for example, O. O'Neil, *Autonomy and Trust in Bioethics*, Cambridge: Cambridge University Press (2002); C. Foster, *Choosing Life, Choosing Death: The Tyranny of Autonomy in Medical Ethics and Law*, Oxford: Hart (2009); G. Laurie, *Genetic Privacy: A Challenge to Medico-Legal Norms*, Cambridge: Cambridge University Press (2002). See Chapter 3 for a deeper discussion of the contemporary significance of the interest in autonomy.

112 See B. Waller, 'The Psychological Structure of Patient Autonomy' *Cambridge Quarterly of Healthcare Ethics* ii (2002), 257, 263 for support for the argument that interference with autonomy constitutes a harm which offends the principle of first do no harm.

how English negligence law might react to the novel genomic claims considered in this book. In general, it might be argued that the English courts currently take a relatively restrictive approach to the question of whether new duties of care should be imposed with respect to new types of harm. The courts might be particularly reluctant to recognize the setbacks which form the essence of the novel claims considered here because they are not universally perceived as deleterious. However, it is clear that the tort of negligence remains capable of recognizing interference with novel interests. This chapter introduces the idea that the interest in autonomy might form the basis of the recognition of harm in the suite of novel genomic negligence claims considered in this book.

Chapter 3

An interest in autonomy as the basis for a new head of damage in negligence

Possible interpretations and limitations

Introduction

For the most part, allowing people to determine the course of their own life is a good thing. An individual who is able to choose freely from understood options, meaningfully directs her own life. The value of autonomy has long been recognized in the philosophical context.[1] It does not have such a long history of legal recognition, but in recent years the principle has become an important bioethical principle, which has influenced its recognition as a fundamental principle in English medical law.[2] With respect to the notion of autonomy as a legal principle, this book seeks to consider the different question of whether autonomy might serve as the basis for a suite of novel genomic negligence claims which might arise from genetic services. In recent years, the House of Lords has begun to tentatively recognize that autonomy is an important personal interest which should receive some legal protection in the event of negligent interference.[3] If negligence law were specifically imbued with a notion of personal autonomy it would have significantly more to offer to victims of novel negligence actions where the harm does not fit with traditional corporeal notions of tortious harm. In the face of the restrictive culture of English tort law described in the previous chapter, this chapter takes its cue from the House of Lords in two fairly recent cases, *Rees v Darlington Memorial NHS Trust*[4] and *Chester v Afshar*,[5] and seeks to present autonomy as a concept that could form the basis of an action in the tort of negligence.

---

1 Two notable eighteenth century autonomy philosophers are John Stuart Mill and Immanuel Kant. More recently see the work of Gerald Dworkin, Harry Frankfurt, Joseph Raz, Robert Young, Richard Lindley and John Christman.
2 *Re B (Consent to Treatment: Capacity)* [2002] EWHC 429; *St George's Healthcare NHS Trust v S, R v Collins Ex p. S (No. 2)* [1999] Fam. 26; *Re T (Adult: Refusal of Treatment)* [1993] Fam. 95.
3 See, in particular, *Rees v Darlington Memorial Hospital NHS Trust* [2004] 1 AC 309 and *Chester v Afshar* [2005] 1 AC 134.
4 *Rees v Darlington Memorial NHS Trust* [2004] AC 309.
5 *Chester v Afshar* [2005] 1 AC 134.

Part I of this chapter analyses the concept of autonomy which is adopted in English medical law, focusing on the fact that it is a liberal interpretation of the concept which does not require specific content. Following on from this, Part II turns to consider how rationality might function as an aspect of autonomy. In this part, the focus is on autonomy as imbued with a notion of rationality, which relates to the individual's ability to impose some objective value on oneself. In this way, rationality is construed here as a substantive concept. Within this context, there is a consideration of how a notion of value, which is central to English negligence law, might be relied on to pour content into rationality which could form the basis of the legal interpretation of autonomy in negligence. Alternatively, Part III considers the concept of rationality from a procedural, as opposed to a substantive or a value, perspective, whereby rationality is determined by one's ability to meet her own subjective ends rather than her ability to impose some objectively valuable ends on herself. These two conceptions of rational autonomy provide the foundation for the novel actions arising from human genetic services, with which this book is concerned.

Finally, Part IV considers whether the value of autonomy is instrumental or intrinsic. This is crucial to the issue of where the boundaries of the novel head of damage might lie. Recognizing the intrinsic value of autonomy will cast the net much wider than recognition of the value as instrumental. However, recognition of the intrinsic notion of autonomy would represent a much truer recognition of the value that lies at the heart of the principle of autonomy.

The concept of autonomy

In presenting a concept of autonomy which might function as a legal interest, this chapter does not attempt to give a comprehensive philosophical account of the principle of autonomy. There are many books which do this.[6] Autonomy is an infinitely wide concept which plays a role in moral,[7] social[8]

---

6 See, for example, G. Dworkin, *The Theory and Practice of Autonomy* Cambridge: Cambridge University Press (1988); H. Frankfurt, *The Importance of What we Care About*, Cambridge: Cambridge University Press (1988); J. Christman, *The Inner Citadel: Essays on Individual Autonomy*, Oxford: Oxford University Press 1989); J. Taylor, *Personal Autonomy*, Cambridge: Cambridge University Press (2005); C. Mackenzie and N. Stoljar (eds), *Relational Autonomy: Feminist Perspectives on Autonomy, Agency and the Social Self*, New York: Oxford University Press (2000); O. O'Neil, *Autonomy and Trust in Bioethics*, Cambridge: Cambridge University Press (2002).

7 C. M. Korsgaard, *The Sources of Normativity*, New York: Cambridge University Press (1996); T. Hill, 'The Importance of Autonomy' in *Women and Moral Theory* E. F. Kittay and D. T Meyers, (eds) New Jersey: Rowman and Littlefield (1987), 129; T. Hill 'The Kantian Conception of Autonomy' in J. Christman, (ed.) *The Inner Citadel: Essays on Individual Autonomy*, New York: Oxford University Press (1989), 91.

8 C. Mackenzie and N. Stoljar (eds), *Relational Autonomy: Feminist Perspectives on Autonomy, Agency and the Social Self*, New York: Oxford University Press (2000); M. Friedman, 'Autonomy and Social Relationships: Rethinking the Feminist Critique' in D. T. Meyers (ed.) *Feminist Rethink the Self*, Boulder, CO: Westview Press (1997), 40.

and political,[9] as well as legal, theory. Given this, there are many things that are not said about theories of autonomy here. Autonomy functions in a variety of contexts. The principle consists of a number of different conceptions, as opposed to one clear unified concept.[10] Thus, when we draw from different kinds of theory we are drawing from different manners of conceptualization.[11] Many conceptions of autonomy are discipline specific and may not be critically relevant to the question of whether autonomy is capable of being recognized as an interest in the tort of negligence.

Thick and thin conceptions of autonomy

The term autonomy is used in an 'exceedingly broad fashion'[12] by different authors who are not referring to the same thing. At a very basic level, it seems that the concept can be described approximately by such terms as 'self-rule, self-determination, self-government or independence'.[13] Thus, central to the principle of autonomy is the notion of a capacity to determine, and de facto determining oneself. In this way, autonomy denotes authorship over one's own life which, in practice, is often associated with the making of one's own choices and decisions. However, it might be argued that a conception of autonomy which focuses on making decisions and choices is a narrow conception. There is a widely held philosophical opinion that autonomy is a much wider concept conceived of as a character ideal, which has a broader application to aspects of our lives.[14] Whilst noting that autonomy is sometimes used as an equivalent to self-rule, Gerald Dworkin argues that it is also equated with dignity, integrity, individuality, independence, responsibility and self-knowledge, and that it pertains to actions, to beliefs, to reasons for acting, to rules, to the will of other persons, to thoughts and to principles.[15] Ultimately, he sees autonomy as

---

9 M. Friedman, *Autonomy, Gender, Politics*, New York: Oxford University Press (2003); A. Ripstein, *Equality, Responsibility, and the Law*, Cambridge: Cambridge University Press (1999).

10 See, for example, S. McLean, *Autonomy, Consent and the Law,* London: Routledge Cavendish (2010), 14; N. Arpaly, *Review Contours of Agency: Essays on Themes from Harry Frankfurt* 113 (2004), *Mind* 452; J. Feinberg, *Harm to Self; The Moral Limits of the Criminal Law*, New York: Oxford University Press (1986), Chapter 18; J. Christman, 'Constructing the Inner Citadel: Recent Work on the Concept of Autonomy' 99 (1988), *Ethics* 109.

11 Thank you to John Coggon for this important point.

12 G. Dworkin, *The Theory and Practice of Autonomy*, Cambridge: Cambridge University Press (1988), 6.

13 J. Feinberg, *Harm to Self; The Moral Limits of the Criminal Law*, New York: Oxford University Press (1986), 27. However, these individualistic conceptions of autonomy might be challenged by theories which seek to promote a communitarian approach to autonomy. See, in particular, C. Mackenzie and N. Stoljar (eds), *Relational Autonomy: Feminist Perspectives on Autonomy, Agency and the Social Self*, New York: Oxford University Press (2000) and W. Gaylin and B. Jennings, *The Perversion of Autonomy*, Washington, DC: Georgetown University Press (2003).

14 R. Young, 'The Value of Autonomy' 32 (1982), *The Philosophical Quarterly* 35, 35; G. Dworkin, *The Theory and Practice of Autonomy*, Cambridge: Cambridge University Press (1988), 6.

15 G. Dworkin, *The Theory and Practice of Autonomy*, Cambridge: Cambridge University Press (1988), 6.

a global notion which refers to states of a person.[16] Although it can also be considered as a local notion which relates to particular traits, motives and values.[17] Christman argues that conceptions of autonomy that see only desires as the focal point are too narrow as people can exhibit autonomy relative to a wide variety of personal characteristics such as values, physical traits and relations to others.[18] Double believes that any element of body, personality or circumstances that figures centrally in reflection and action should be open to appraisal in terms of autonomy.[19] Ronald Dworkin adopts a similarly wide interpretation of the concept of autonomy as 'the ability to act out of genuine preference or character or conviction or a sense of self'.[20] The notable volume of varying interpretations of autonomy has led Gerald Dworkin to conclude that the only feature of autonomy which is constant across theories is that 'autonomy is a feature of persons and that it is a desirable quality to have'.[21]

Christman argues that at its most basic level of application, autonomy is properly seen as the property of preferences or desires (or their formation), while another view is that it is a property of whole persons or of persons' whole lives.[22] Young argues that this latter image of autonomy as a global notion, which relates to the unified ordering of a person's life, is what exemplifies most fully its self-directedness. On this analysis, autonomy is not simply about exercising choice; it pertains to a wider, global concept of what it means to be one's own author. In the legal sense the question of whether a person has the capacity to be and is, in fact, autonomous, usually arises in the context of individual decision-making. In a more general sense, autonomy is a principle which has a wider application in establishing conditions under which people can freely make up their own minds about what to believe and how to live, and can then act accordingly.[23] On this latter conception, autonomy is not simply a feature of persons which only requires respect in relation to specific decisions which an individual makes, it is a comprehensive feature of persons which requires more than respecting others' explicit choices. Young advocates a particularly wide interpretation of autonomy and speaks of coerced individuals hanging on to autonomy at the level of their thoughts,[24] despite the fact that many would perceive the

16 Ibid. 13–14 and 19–20.

17 J. Christman, 'Autonomy in Moral and Political Philosophy', Stanford Encyclopedia of Philosophy (2009). Available HTTP: http://plato.stanford.edu/entries/autonomy-moral/ (accessed 8 September 2009).

18 Ibid.

19 R. Double, 'Two Types of Autonomy Accounts' 22 (1992), *Canadian Journal of Philosophy* 65, 66.

20 R. Dworkin, *Life's Dominion*, London: Harper Collins (1993), 225.

21 G. Dworkin, *The Theory and Practice of Autonomy*, Cambridge: Cambridge University Press (1988), 6.

22 J. Christman, 'Constructing the Inner Citadel: Recent Work on the Concept of Autonomy' 99 (1988), *Ethics* 109, 111.

23 A. Voorhoeve, 'The Limits of Autonomy' 46 (2009), *The Philosophers Magazine* 78.

24 R. Young, 'The Value of Autonomy' 32 (1982), *The Philosophical Quarterly* 35, 38.

circumstances of the individual described by Young; that is living in 'severe repression and coercion',[25] as a perfect example of a person whose autonomy is prevented.

A wider interpretation of autonomy provides greater scope for portraying autonomy as a principle which is intimately connected with other values. In *On Liberty*, Mill claimed that autonomy is 'one of the elements of well being'.[26] In essence, this links the notion of well being with self-determination. Raz, in particular,[27] considers that the free choice of goals and relations is an essential ingredient of well being. On this perspective, well being is determined by success in willingly endorsed pursuits.[28] Indeed, Raz goes as far as saying: 'In western industrial societies a particular conception of individual well-being has acquired considerable popularity. It is the ideal of personal autonomy.'[29]

Portraying autonomy as such a global conception of the individual good life demonstrates the potential breadth of the concept.[30] If the courts were to adopt a conception of autonomy as the basis for recognition of a novel kind of harm in negligence, they may be unlikely to adopt a wide conception which they would have difficulty defining and limiting.[31] Furthermore, in the context of a negligence action, the interference with autonomy would have to arise from a culpable action which, in the context of the genomic claims considered in this book, is largely an interference with a person's explicit decision in relation to some issue. Throughout this book the potential claim is based on the claimant's express failure to embrace and endorse the circumstances in which she finds herself. In each of the cases those circumstances were created by another who was aware that the particular outcome was not desired. Largely, this is because the

25 Ibid. 38.

26 J. S. Mill, *On Liberty* 1859 in D. Spitz (ed.), New York: Norton (1975), Chapter 3.

27 J. Raz, *The Morality of Freedom*, New York: Oxford University Press (1986), Chapter 14. This chapter does not seek to consider what well being amounts to; the aim is simply to highlight its potential link to theories of autonomy.

28 Ibid. 369.

29 Ibid. Chapter 14, in particular, 369.

30 It also highlights the difficulty of determining where the value lies in autonomy. Is it a principle which ought to be respected because it allows the individual to achieve that which contents her and ensures her well being? Or is there something intrinsic in protecting well being that is contingent on us experiencing being treated as autonomous, irrespective of whether we achieve what we wanted to achieve through the exercise of autonomy? The issue of whether autonomy is an intrinsic or an instrumental value is considered in detail below and in Chapter 5.

31 For the courts want to be able to control the development of novel legal concepts through definition and conceptualization see *Wainwright v Home Office* [2002] QB 1334, Mummery J, 1351. Although, as the discussion below demonstrates, the courts adopt a liberal interpretation of autonomy which is largely devoid of content and therefore not easily limitable in the context of the tort of trespass to the person in relation to consent to medical treatment.

individual has expressed her desire not to be put in those circumstances.[32] However, this book also considers scenarios where the individual finds herself in circumstances which she does not embrace, but which were not expressly rejected by her. Here it is argued that the fact that she did not have the opportunity to express her preferences does not mean that her autonomy cannot have been interfered with. Indeed, some tort scholars agree that negligence law's protective purpose can generally be seen as a commitment to the ideal of personal autonomy,[33] suggesting that prior stipulation of one's choice or desire not to be treated negligently is not crucial to the question of whether that negligence interferes with her autonomy. An individual might endorse her circumstances indicating that they do not interfere with her autonomy, but if she does not endorse them, they potentially amount to an interference with her autonomy. Royce embraces this notion of endorsement in determining whether a person's desires can be said to be her own. He comments: 'His devotion is his own. He chooses, it, or, at all events, approves it'.[34]

Some actions could largely be construed as interferences with autonomy because it is known that the person will not willingly endorse them even though she does not expressly say so. Inflicting bodily harm, rape or stealing might be seen to interfere with a person's autonomy even though she is not given a choice about whether these events occur. The same can be said for those losses that might be inflicted through negligence, which English negligence law currently recognizes, but which are not based on prior express decisions. The conception of autonomy, which is engaged in this book, is a wider conception which is not contingent on expressed decisions, but reflects the interpretation of the concept as a wider feature of persons.[35]

Autonomy in English medical law

The aim in this chapter is to focus on conceptions of autonomy which are capable of forming the basis of a legal interest in the tort of negligence. Thus, autonomy is discussed here as imbued with substantive or value rationality and procedural rationality. These interpretations are defended from the perspective of their practical ability to function as recognizable

32 In practice, there may be reasons for overriding one's autonomy where to respect it would conflict with the interests of others. However, this does not mean one's choice was not autonomous, rather that there is justification for not respecting one's autonomous choice.
33 R. Mullender, 'English Negligence Law as a Human Practice' 21 (2009), *Law and Literature* 321, 324–325.
34 J. Royce, *The Philosophy of Loyalty*, New York: Macmillan (1908), 21.
35 To attempt to reflect this wider interpretation of autonomy, this book largely speaks of autonomy in reaction to choices, actions and desires. It is hoped that this reflects the argument expressed here that autonomy is widely conceived as something more than a decisional concept.

harm in the tort of negligence. The underlying rationale is that autonomy is a desirable quality to have and that some protection of it is better than none. However, practically speaking, this discussion is underpinned by a recognition that the English negligence system cannot protect against every interference with self-authorship. On this basis, if negligence law recognized interference with the interest in autonomy as a basis for legal harm, it might be inclined to reject an individualistic account of autonomy which has come to mean the satisfaction of individual wishes and desires,[36] in favour of a concept of autonomy which is imbued with normatively substantive conditions, such as the ability to recognize and follow certain moral or political norms.[37] Feminist scholars adopt a similar approach whereby the recognition of certain basic value claims is a requirement of autonomy.[38]

Despite making some attempt to recognize the harm caused by negligent interference with personal autonomy on two occasions,[39] the House of Lords has failed to address the issue of what personal autonomy actually consists in. Furthermore, in the medical law context, where the concept of autonomy is significantly more developed than it is in the tort of negligence, the courts have not defined precisely what autonomy is. As Coggon notes: 'It is rare for judges to provide an explicit, philosophical investigation of autonomy'.[40]

Despite this, a recurring theme with respect to the definition of autonomy in the context of consent to medical treatment has been whether rationality is a condition of autonomy.[41] Thus, legally speaking, the issue of whether autonomy is conditional focuses on whether it is imbued with an aspect of rationality. Nevertheless, the courts have no more defined rationality than they have defined autonomy, and like autonomy, the concept of rationality is subject to varying interpretations.[42] Taking the cue from English medical law that there is an important question to be asked about whether rationality is an aspect of autonomy, this chapter

---

36 S. Mclean, *Autonomy, Consent and the Law*, London: Routledge-Cavendish (2010), 15.

37 P. Benson, 'Freedom and Value' 84 (1987), *Journal of Philosophy* 465 and S. Wolf, *Freedom and Reason*, New York: Oxford University Press (1990).

38 M. Oshana, 'Personal Autonomy and Society' 29 (1998), *Journal of Social Philosophy* 81; N. Stoljar, 'Autonomy and the Feminist Intuition' in C. Mackenzie and N. Stoljar (eds), *Relational Autonomy: Feminist Perspectives on Autonomy, Agency and the Social Self*, New York: Oxford University Press (2000), 94.

39 *Rees v Darlington Memorial Hospital NHS Trust* [2004] 1 AC 309 and *Chester v Afshar* [2005] 1 AC 134.

40 J. Coggon, 'Varied and Principled Understandings of Autonomy in English Law: Justifiable Inconsistency or Blinkered moralism?' 15 (2007), *Health Care Analysis* 235, 236.

41 *Re T (Adult: Refusal of Treatment)* [1993] Fam 95, Lord Donaldson 102; *Re MB*, Butler-Sloss.

42 See, in particular, L. Koch, 'IVF – An Irrational Choice?' 3 (1990), *Journal of International Feminist Analysis* 235.

considers how the concept of autonomy might function as a legally recognized interest in the tort of negligence if it were imbued with a substantive value or a procedural notion of rationality. Before this, however, let us consider how rationality has, in fact, been interpreted with regard to the principle of autonomy in the context of English medical law.

Currently, from a medical law perspective, it seems that what is perceived to be good about autonomy is its capacity to permit the individual to make her own decisions free from external constraints. This is evidenced by the regular assertion that an adult patient who suffers from no mental incapacity has an absolute right to choose whether to consent to medical treatment, to refuse it or to choose one rather than another of the treatments being offered.[43] In essence English medical law adheres to an individualistic conception of autonomy which appears to be linked to its need to shed the paternalistic past of medicine rather than reflect on the metaphysical source of what it is to be autonomous.[44] The courts have repeatedly stated that this right of choice is not limited to decisions which others might regard as sensible. It exists notwithstanding that the reasons for making the choice are rational, irrational, unknown or even non-existent.[45] In this sense, a decision which is procedurally independent[46] and made by one with capacity will ostensibly be recognized as autonomous. However, where the theory of autonomy is imbued with an aspect of rationality, the question of whether a decision is autonomous is different. Rationally autonomous choice relies on the value of one's ends; the notion of value being referenced to some external conception. Alternatively, rationality might depend upon one's procedural ability to do what she truly wants to do; that is the ability to make her authentic desires effective in action.[47]

It is within the context of informed consent that the principle of autonomy has established itself as an interest that is capable of being protected by English tort law. Through the tort of trespass to the person, as opposed to

---

43 *Re T (Adult: Refusal of Treatment)* [1993] Fam 95.

44 S. McLean, *Autonomy, Consent and the Law*, London: Routledge Cavendish (2010), 17. McLean relies on the alleged distinction between between the individualistic and relational models of autonomy to argue that, in the context of consent, although the law purports to prefer the individualistic model, it vacillates between the two to achieve policy-based objectives.

45 This premise has also been acknowledged in a number of other cases. For example, *Sidaway v Board of Governors of the Bethlem Royal Hospital and the Maudsley Hospital* [1985] AC 871, Lord Templeman, 904–905; *Re MB (Medical Treatment)* [1997] 2 FLR 426, Butler-Sloss LJ, 432; *Re T (Adult: Refusal of Treatment)* [1993] Fam 95, Lord Donaldson, 102; *Re B (Consent to Treatment: Capacity)* [2002] 1 FLR 1090, Dame Butler-Sloss, 1095.

46 In the sense that it is not unduly influenced or made as a result of coercion or manipulation by others.

47 As McLean notes whilst it might be relatively easy to identify the autonomous person, it is much less easy to decide whether a decision is truly autonomous. S. McLean, *Autonomy, Consent and the Law*, London: Routledge Cavendish (2010), 69.

negligence,[48] the courts have sought to protect deliberate interferences with patients' medical decisions. Nevertheless, despite the court's ostensible adherence to an individualistic account of autonomy,[49] and express denials that rationality is an aspect of autonomy,[50] it might be argued that they *do* in fact require the patient's decision to be rational on some level. First, autonomy in the context of medical decision-making relates to the making of a decision from a range of options. The existence of specific options may establish the rationality of the decision which the patient can actually make.[51] In this way, there is an objectively rational framework within which patients can decide. However, the decision to refuse all treatment cannot be taken away from the patient[52] and it is in the context of ostensibly unwise decisions to refuse treatment that the question of patient autonomy usually comes before the court.[53] Where an ostensibly competent person makes an apparently irrational decision to refuse treatment, the courts may seek to override the decision on the grounds that the person is incapacitated and therefore incapable of autonomy, as opposed to acknowledging that they require patients to be able to give rational reasons for their decisions.[54]

One of the strongest reassertions of the principle of autonomy was by Butler-Sloss in *Re MB*. A pregnant woman refused to go ahead with a caesarean section solely because the insertion of a needle was necessary to perform the operation. Butler-Sloss said:

> A competent woman who has the capacity to decide may, for religious reasons, other reasons, for rational or irrational reasons or for no reason at all, choose not to have medical intervention, even though the consequence may be the death or serious handicap of the child she bears, or her own death. In that event the courts do not have the jurisdiction to declare medical intervention lawful and the question of her own best interests, objectively considered, does not arise.[55]

48 See, in particular, *Re B (Consent to Treatment: Capacity)* [2002] 1 FLR 1090.

49 S. McLean, *Autonomy, Consent and the Law*, London: Routledge Cavendish (2010).

50 *Re MB (Medical Treatment)* [1997] 2 FLR 426, Butler-Sloss LJ, 432; *Re T (Adult: Refusal of Treatment)* [1993] Fam 95, Lord Donaldson 102; *Re B (Consent to Treatment: Capacity)* [2002] 1 FLR 1090, Dame Butler-Sloss, 1095.

51 That is what the medically qualified person believes to be a sensible treatment option in the circumstances. My thanks go to John Coggon for pointing this out.

52 As the decision to access certain treatments might. Although this was not the essence of the decision in *Burke*, it seems to be a necessary implication of it.

53 *Re MB (Medical Treatment)* [1997] 2 FLR 426, Butler-Sloss LJ, 432; *Re T (Adult: Refusal of Treatment)* [1993] Fam 95, Lord Donaldson, 102; *Re B (Consent to Treatment: Capacity)* [2002] 1 FLR 1090, Dame Butler-Sloss, 1095; *Re C (Adult: Refusal of Medical Treatment)* [1997] 1 WLR 290.

54 Either in the sense of value rationality or that the patient is able to demonstrate that the (unwise) decision represents her authentic desires.

55 *Re MB (Medical Treatment)* [1997] 2 FLR 426, Butler-Sloss LJ, 437.

She explained that an irrational decision is one which is so outrageous in its defiance of logic or of accepted moral standards that no sensible person who had applied his mind to the question to be decided could have arrived at it. Despite this dictum, Butler-Sloss concluded that at the moment of panic MB's fear dominated all and at that point she was not capable of making a decision.[56] As noted above, capacity is considered to be a necessary condition for autonomy. Nevertheless, there was no clear evidence that MB failed to reach the requirements for capacity; namely an ability to comprehend and retain treatment information, believe it and weigh it in the balance to arrive at a choice.[57] In effect, the courts did override MB's decision on the basis that it was irrational. Indeed, in a note to the court, the consultant psychiatrist who had assessed MB acknowledged that the central problem with MB's refusal of treatment was its irrationality. He said: 'It was her irrational fear of needles that has got in the way of proceeding with the operation.'[58]

But as Butler-Sloss noted, although it might be thought that irrationality sits uneasily with competence to decide, irrationality in itself does not as such amount to incompetence.[59]

This was clearly a very difficult case and it may be that MB did lack capacity. The problem was that no serious attempt to assess her capacity within the context of the prevailing legal standard was evident. However, MB's decision could be deemed irrational from the perspective of objective notions of rationality and from the procedural perspective of being an effective exercise of her own authentic desires. A decision which would lead to the still birth of a healthy and wanted viable baby would not be seen as rational by most people. Moreover, the refusal of the needle was not consistent with MB's own stated desire to have the caesarean section and deliver a live, healthy baby.[60] The suspicion is that the court's belief that MB's decision was irrational did, in fact, lead to the decision to find her incapable of autonomy despite the court's explicit statement that an autonomous decision need not be rational. If autonomy is the individualistic construct the court portrays it as, then autonomy is a minimally conditional concept, whereby a competent person's choice, action or desire is respected simply because it is hers and nothing more.[61] However, when patients make seemingly odd decisions; namely those to reject medical advice leading to death, the courts

---

56 Ibid. Butler-Sloss LJ, 438.
57 This test was known as the *Re C* test because it originated from *Re C (An Adult: Refusal of Treatment)* [1994] 2 FCR 151, Thorpe J, 156. It was the dominant test for capacity at the time, but has now been superseded by similar requirements in s. 3 of the Mental Capacity Act 2005.
58 *Re MB* [1997] 2 FLR 426, Butler-Sloss LJ, 438.
59 Ibid. Butler-Sloss LJ, 437.
60 Ibid. Butler-Sloss LJ, 427.
61 Provided the person is capacitated and free from interferences which obstruct procedural independence, such as coercion or manipulation.

do sometimes appear to require that the patient demonstrates the rationality of her choice, action or desire before it is deemed to be autonomous. Where a patient's choice, action or desire appears odd, it seems that she must demonstrate the rationality of it by presenting the court with convincing reasons for it.[62]

Thus, despite the numerous expressions to the contrary, it seems that the courts will investigate the rationality of odd choices before they deem them to be autonomous and, therefore, worthy of respect. Despite this, the courts persistent denial that rationality is a requirement of autonomy has created a general perception that rationality is not, in fact, an aspect of the concept of autonomy in English medical law. Given this, there is little in the way of analysis of the nature of the concept of rationality that might underpin legal autonomy.[63] Thus, despite the fact that, explicitly, the law subscribes to a basic, liberal conception of autonomy which is not based on particular values and is therefore content-neutral,[64] in reality the courts subscribe to an ideal version of autonomy, of which rationality *is* an aspect.

Autonomy and recognition of novel interests in the tort of negligence

The tort of negligence does not provide protection in the event of all negligent interferences with individuals. Indeed, where negligence has evolved to protect novel interests, protection is limited. For example, in fairly recent years negligence has developed to protect the interest in not being mentally harmed as a result of another's negligence. However, this interest will only be legally recognized in tightly circumscribed circumstances, namely where the harm reaches the threshold of pathological. Furthermore, limitations are also applied based on the reasonableness of the fear leading to the harm,[65] the relationship to the person suffering the harm, the medium of the discovery of the event leading to the harm and the time elapsed between the occurrence of the event and the individual's discovery of it.[66] Although the courts have recognized the interest

---

62 In saying this, it is important to note that the courts have a role in protecting vulnerable patients whose capacity for autonomy is in doubt. In this respect they may often have to make some deeper examinations of the patient's choice, action or desire to determine whether she has capacity.

63 For an interesting account of how rationality might be conceived to be an aspect of autonomous decision-making in the context of medical law see J. Savulescu and R. W. Momeyer, 'Should Informed Consent be Based on Rational Beliefs?' 23 (1997), *Journal of Medical Ethics* 282.

64 Autonomy in this sense is conceived as the ability to be self-directed, free from procedural independence and a choice, action or desire can be autonomous no matter how depraved or morally valueless it is. J. Christman, 'Autonomy in Moral and Political Philosophy', Stanford Encyclopedia of Philosophy (2009). Available HTTP: http://plato.stanford.edu/entries/autonomy-moral/ (accessed 8 September 2009).

65 See *Page v Smith* [1996] AC 155.

66 See *Alcock v Chief Constable of South Yorkshire Police* [1992] 1 AC 310. Damages for economic loss are also limited.

in not being harmed in the mental sense, they do not see it as of significant priority as protection for the interest in being physically harmed.

From the perspective of the evolution of the action for psychiatric harm in negligence, it might be argued that the English courts would be more willing to recognize that a novel interest, namely autonomy, might form the basis for legally recognized harm where the interest is amenable to limitation by the courts.[67] An ideal conception of autonomy as imbued with rationality, as opposed to a liberal conception which rests simply on capacity and independence, would allow greater analytical purchase on what autonomy actually consists in, thereby providing potential boundaries to the legal recognition of the interest. In this way, the courts would be able to prioritize those interferences with autonomy which they perceive to be the most deserving of legal recognition.

Indeed, the law of negligence represents a system of protecting ideals as opposed to protecting holistic basic values. This is demonstrated by the limited ability of negligence to provide comprehensive protection for those interests which it currently recognizes. On this basis, the underpinning rationale for this chapter is that a workable legal notion of autonomy could be based on an ideal notion of autonomy which is imbued with a recognition of rationality. Whilst the basic liberal conception of autonomy rests on the notion of value neutrality, some writers insist that the autonomous person must enjoy substantive, as well as procedural, independence.[68] Theories of autonomy that rest on substantive independence stipulate the 'content of the desires and values and so on, in virtue of which one is considered autonomous'.[69]

---

67 For the courts desire to be able to control the development of novel legal concepts through definition and conceptualization, see *Wainwright v Home Office* [2002] QB 1334, Mummery J, 1351.

68 N. Stoljar, 'Autonomy and the Feminist Intuition' in C. Mackenzie and N. Stoljar (eds), *Relational Autonomy: Feminist Perspectives on Autonomy, Agency and the Social Self*, New York: Oxford University Press (2000), 94; P. Benson, 'Freedom and Value' 84 (1987), *Journal of Philosophy* 465; P. Benson, 'Feminist Intuitions and the Normative Substance of Autonomy' in J. S. Taylor (ed.), *Personal Autonomy: New Essays on Personal Autonomy and Its Role in Contemporary Moral Philosophy*, Cambridge: Cambridge University Press (2005), 124 and 2005; M. Oshana, *Personal Autonomy in Society*, Hampshire: Ashgate (2006). The ideal of autonomy as self-sufficient individualism is criticized by C. Mackenzie and N. Stoljar, 'Autonomy Reconfigured' in C. Mackenzie and N. Stoljar (eds), *Relational Autonomy: Feminist Perspectives on Autonomy, Agency and the Social Self*, New York: Oxford University Press (2000), 25. Many feminists are critical of this individualistic interpretation of autonomy which they see as unrealistic and ignorant of the essentially social nature of people. Indeed, MacKenzie and Stoljar argue that the concept of autonomy is conflated with one particular conception of autonomy; the self-sufficient, rugged male individualist, rational maximizing chooser of libertarian theory. The collection of feminist essays by MacKenzie and Stoljar seeks to reconfigure autonomy in a way which reflects the social and relational aspects of persons. For a further attack on the individualistic notion of autonomy, see W. Gaylin and B. Jennings, *The Perversion of Autonomy*, Washington, DC: Georgetown University Press (2003).

69 J. Christman, 'Autonomy in Moral and Political Philosophy', Stanford Encyclopedia of Philosophy (2009). Available HTTP: http://plato.stanford.edu/entries/autonomy-moral/ (accessed 8 September 2009).

Christman argues that on such a substantive conception, the concept of auton-omy is reserved only to those whose lifestyles and value pursuits are seen as acceptable from some political or theoretical point of view.[70] Relational and moral interpretations of autonomy might fall into this category in that, to be deemed to be autonomous, a particular choice, desire or action must accord with some external, objective conditions.[71] This type of autonomy effectively refers to the ability to govern oneself according to external (moral) principles.[72] On the other hand, autonomy might be conceived as a concept which depends on procedural, as opposed to substantive, conditions, whereby the individual must be capable of critically reflecting upon her desires to ensure that those desires which are effective in action reflect her true or authentic self.[73]

This book focuses on how the concept of rationality can be interpreted sub-stantively or procedurally to present a conditional account of autonomy. In this way, the concept of rationality is employed as an evaluative tool to enable an analysis of how autonomy might function as a conditional concept as the basis for the recognition of a novel interest in the tort of negligence. It is hoped that focusing on the notion of rational autonomy provides an autonomy perspective which is different to the one which the courts have explicitly adopted with respect to medical trespass actions. It might be argued that focusing on auton-omy as imbued with rationality presents a more realistic focus for the *legal* recognition of autonomy, thereby providing a workable basis for the recognition of the interest in autonomy in the tort of negligence. Before making the substantial move of recognizing a new type of damage, history suggests that the courts might want to be able to foresee where its potential limits may lie. Thus, whatever the advantages and disadvantages of requiring rationality as an aspect of autonomy in an abstract theoretical sense, there is merit in arguing for its requirement as part of a legal conception of autonomy because it allows some objective evaluation of what autonomy consists in, which makes legal recognition of the interest more likely.

Rationality as an aspect of autonomy

It might be that there are many notions which could be relied on to imbue the principle of autonomy with particular values or procedural standards, which would define autonomy as more than simply doing as one pleases. However, the question of whether rationality is an aspect of autonomy has been debated by philosophers since Kant proposed a particularly intimate

---

70 Ibid.
71 N. Stoljar, 'Autonomy and the Feminist Intuition' in C. Mackenzie and N. Stoljar (eds), *Relational Autonomy: Feminist Perspectives on Autonomy, Agency and the Social Self*, New York: Oxford University Press (2000), 94; M. Oshana, *Personal Autonomy in Society*, Hampshire: Ashgate (2006).
72 Kant is perhaps the most famous proponent of such a version of autonomy.
73 This consideration of this perspective in this work draws heavily on the work of Gerald Dworkin.

relationship between the two concepts. Furthermore, as the above discussion demonstrates, in the context of medical law, the question of whether rationality ought to be an aspect of autonomy has also repeatedly arisen. Given this, this chapter focuses on how a principle of autonomy, which is imbued with a rationality condition, might function as the basis for the recognition of a novel interest in negligence.[74]

Philosophers are divided on the question of whether the concept of rationality is an aspect of autonomy. In a review of the literature on the concept of rationality as an aspect of autonomy, Christman concludes that a consensus has not been reached on the question of whether an agent must be rational to be autonomous.[75] Some commentators believe that a contradiction arises in the notion of *self*-rule, which is at the heart of the individualistic concept of autonomy if rationality is required. This is because applying a condition to the process of autonomous decision-making necessarily opens up that process to scrutiny. Berlin argues:

> Once I take the view (that freedom requires that I act in accordance with the demands of reason) I am in a position to ignore the actual wishes of men, to bully, oppress, torture them in the name ... of their 'real' selves in the secure knowledge that whatever is the true goal of man, must be identical with his freedom – the free choice of his 'true', albeit often submerged and inarticulate self.[76]

Christman suggests that requiring rationality for autonomy essentially conflicts with the core idea of *self*-government that autonomy is meant to express.[77] He adds:

> any model of autonomy that demands that the autonomous agent be rational will, by that token, imply that agents with varying degrees of decision-making competence will then possess the property of autonomy to varying degrees.[78]

Young takes a more nuanced approach to the question of whether rationality ought to be an aspect of autonomy. He concedes that not being seriously irrational does seem to be necessary for autonomy. However, he continues:

---

74 What follows is not an attempt to provide a comprehensive account of rationality, but rather some specific ways that the concept of rationality might be conceived as comprising an aspect of autonomy.

75 J. Christman, 'Constructing the Inner Citadel: recent Work on the Concept of Autonomy' 99 (1988), *Ethics* 109, 116.

76 I. Berlin, 'Two Concepts of Liberty' in *Four Essays on Liberty*, London: Oxford University Press (1969), 133.

77 J. Christman, 'Constructing the Inner Citadel: recent Work on the Concept of Autonomy' 99 (1988), *Ethics* 109, 116.

78 Ibid.

we must be wary of ... over-emphasising the role of rationality either by making logical calculation too big a part of the cognitive or by stressing the cognitive at the expense of the affective ... for most of us at some time, acceding to the demands of the rational is liable to restrict our occurent autonomy and perhaps thereby our global autonomy by precluding adventurousness and bolstering inhibition.[79]

Individualists find the notion that autonomy should be subject to any content, over and above procedural independence, which might enable external evaluation of a person's autonomy, inimical. Nietzsche, in particular, adopted a stringent individualistic interpretation of autonomy whereby autonomy and morality are mutually exclusive.[80] For Nietzsche, autonomy required freedom from all external constraints, including the limits of moral obligation, on one's behaviour. From this perspective, the autonomous individual is a supramoral being who has his own independent protracted will, which can never be universalized.[81] Nietzsche believed that the concept of universalizability trained the individual to be a function of the herd, fostering conformity and obedience.[82] Subjecting one's will to a universal moral law limited one's possibilities in life and stifled new ways of thinking. In Nietzsche's view, people should define themselves by their own value and disregard the values of the herd.[83] In this way, Nietzsche was an extreme individualist.[84] Unsurprisingly, given Nietzsche's view of autonomy as devoid of moral foundation, he was contemptuous of the notion of rationality as a vehicle for ordering life. In *Twilight of the Idols* he stated: 'Honest things, like honest men do not carry their reasons exposed. It is indecent to display all one's goods. What first has to have itself proved is of little value.'[85]

79 R. Young, 'Autonomy and Socialization' 356 (1980), *Mind* 567.

80 F. Nietzsche, *On the Genealogy of Morals*, translated by W. Kaufmann and R. J. Hollingdale in *On the Genealogy of Morals and Ecce Homo*, New York: Random House (1967). G. Dworkin, *The Theory and Practice of Autonomy*, Cambridge: Cambridge University Press (1988), 12.

81 F. Nietzsche, *On the Genealogy of Morals*, translated by W. Kaufmann and R. J. Hollingdale in *On the Genealogy of Morals and Ecce Homo*. New York: Random House (1967).

82 F. Nietzsche, *The Gay Science*, translated by W. Kaufmann, New York: Random House (1974).

83 Ibid.

84 J. P. Stern, *Nietzsche*, Glasgow: Collins (1978), and H. Kariel, 'Nietzsche's Preface to Constitutionalism' 25 (1963), *Journal of Politics* 211; P. Carus, *Nietzsche and Other Proponents of Individualism*, Chicago: Open Court Publishing (1914); E. Schuré, 'Individualism and Anarchy in Literature' (2009) 40, *Philosophical Forum* 181. Others deny that Nietzsche values individualism above all else. See, for example, J. Young, *Nietzsche's Philosophy of Religion*, Cambridge: Cambridge University Press (2006); K. Ansell-Pearson, 'Nietzsche on Autonomy and Morality: The Challenge to Political Theory' 39 (1991), *Political Science* 270. In particular, 276.

85 F. Nietzsche, *Twilight of the Idols*, translated by W. Kaufmann in *The Portable Nietzsche*, New York: Viking Press (1968), 41.

More recently Hoffmaster argues that individualism is to be valued because it permits the avoidance of vulnerability and requires self-sufficiency.[86] From this individualistic perspective, the autonomous self is not subservient to some external authority such as a universal moral law. Thus, there is no requirement that the will possesses particular content; autonomy boils down to simply doing as one pleases.[87] Individualists promote the exercise of one's choices, actions and desires and oppose external interference with these whether by society, or any other group or institution. Contemporary Western society seems to reflect this value perspective by placing significant value on the notion of individualism as opposed to collectivism.[88]

Others argue that the preoccupation with individualistic self-determination might be nothing more than a widespread and fashionable obsession, lacking in rational justification.[89] Autonomy is thought to be central in certain moral frameworks, both as a model of the moral person and as the aspect of persons which ground others' obligations to them.[90] For Kant, self imposition of the moral law is autonomy.[91] He believed that all judgments on the value of ends must be subordinated to the obligatory universality of the moral law, derived from the concept of rationality itself.[92] O'Neil argues that, for Kant, the idea of autonomy captures the two central aspects of his account of practical reason; that duty is a matter of acting on principles or laws and those principle or laws should not be derived from

---

86 B. Hoffmaster, 'What Does Vulnerability Mean?' 36 (2006) *Hastings Center Report* 38, 42. For further support see B. Waller, 'Responsibility and Health' 14 (2005) *Cambridge Quarterly of Healthcare Ethics* 177.

87 This is an account of the individualistic nature of Nietzsche's conception of autonomy, rather than an attempt at a comprehensive analysis of Nietschean theory. Thus, it does not consider the robust account of character that underpins Nietzsche's view that one acts freely when one's actions stem from one's own character. He argues that human beings have opposite and not merely opposite drives and value standards that fight each other and rarely permit each other any rest. F. Nietzsche, *Beyond Good and Evil*, translated by W. Kaufmann, New York: Random House (1966), 200.

88 A study by Geert Hofstede of 53 countries across five continents found that Individualism prevails in developed and Western countries, while Collectivism prevails in less developed and Eastern countries. Available HTTP: http://stuwww.uvt.nl/~csmeets/PAGE3.HTM. (accessed 5 January 2010); See also C. Mackenzie and N. Stoljar, 'Autonomy Reconfigured' in C. Mackenzie and N. Stoljar (eds), *Relational Autonomy: Feminist Perspectives on Autonomy, Agency and the Social Self*, New York: Oxford University Press (2000), 25.

89 S. Kristinsson, 'Autonomy and Informed Consent' 10 (2007), *Medicine, Health Care and Philosophy* 253, 255.

90 J. Christman, 'Autonomy in Moral and Political Philosophy', Stanford Encyclopedia of Philosophy (2009). Available HTTP: http://plato.stanford.edu/entries/autonomy-moral/ (accessed 8 September 2009).

91 I. Kant, *Groundwork of the Metaphysics of Morals*, translated by M. J. Gregor in *Kant's Practical Philosophy*, Cambridge: Cambridge University Press (1996), 439.

92 P. Guyer, Kant, Immanuel in E. Craig (ed.), *Routledge Encyclopedia of Philosophy*, London: Routledge. Available HTTP: www.rep.routledge.com/article/DB047 (accessed 22 June 2010).

some arbitrary standing point.[93] For Kant to be autonomous 'is emphatically not to be able to do or have whatever one desires, but rather it is to have the capacity of rational self-governance'.[94] Thus, a Kantian account of rationality rests on the moral law which is self imposed, but its content is not premised on individual circumstances or desires.[95] In this way, Kant formulates a substantive account of rationality. Substantive interpretations of rationality are inimical to individualistic perceptions of the principle of autonomy because if we are to make rational choices, we are therefore governed by the rules of rationality that are not themselves the products of our choices.

The work of Immanuel Kant remains one of the most recognized sources of our understanding of the ethical significance of the concept of autonomy. Although there is little agreement about how Kant conceived autonomy, it is clear that rationality was a central element. Guyer interprets Kant's conception of autonomy as 'the property of the will by which it is a law to itself, or since law must be universal, the condition of an agent who is subject only to laws given by himself but still universal'.[96] On this analysis, the achievement of autonomy is characterized by the ability of the self to transcend its dual nature of a creature of instinct (heteronomy) and a creature of reason (autonomy) and achieve a level of rationality where it can freely prescribe to itself laws which are rational because they are valid for all rational beings.[97] On this view, to be autonomous is to be rational in that one is motivated by reasons uncorrupted by desire and not particular to one's circumstances; i.e. things that would prevent one's maxim being willed universally. Reasons are central to rationality and although we are subject to the temptations of desire, we allow reason rather than these urges to guide our actions.

From this perspective, autonomy has an objective basis. The objective requirement finds form in the categorical imperative which requires that we only act in such a way that we can will that our maxims become universal laws. Hobbes agrees that rationality can be construed substantively. Hobbes argues that to be rational, action must be in line with the *nomos* of the decision-maker.[98] This means that the 'ends pursued relate to some overarching norms and principles, rationally subscribed to by the actor, which take priority over

---

93 O. O'Neil, 'Kant, Rationality as Practical Reason', *Oxford Handbook of Rationality*, A. J. Mele and P. Rawling (eds), Oxford: Oxford University Press (2004), 93–109.

94 R. Downie, J Macnaughton, *Bioethics and the Humanities: Attitudes and Perceptions*, London: Routledge-Cavendish (2007) 42.

95 As McLean notes the Kantian conceptualisation of autonomy is at odds with the way autonomy is used in law to mean the satisfaction of individual wishes and desires. S. McLean, *Autonomy, Consent and the Law,* London: Routledge Cavendish (2010), 15.

96 P. Guyer, 'Kant on the Theory and Practice of Autonomy' 20 (2003), *Social Philosophy and Policy* 70, 70.

97 K. Ansell-Pearson, 'Nietzsche on Autonomy and Morality: The Challenge to Political Theory' 39 (1991), *Political Studies* 270, 273.

98 D. van Mill, 'Rationality, Action and Autonomy in Hobbes's *Leviathan*' XXVII (1994), *Polity* 297.

more immediate interests'.[99] The *nomos* for Hobbes's rational person is over-whelmingly the path of peace.[100] For Hobbes, rational thoughts and actions are guided by the natural law. He believed the laws of nature provided the guidance that allows us to promote the key primary goods upon which our broad life plan should be based.[101] The implication of such substantive or value-based conceptions of autonomy is that we are only acting autono-mously when we freely opt to do the right thing for the right reason. For Kant, the right thing is determined by its universalizability. For Hobbes, it is determined by reference to the natural law. But in the absence of a par-ticular theory of value, it is not clear how the right thing should be determined.

Kant's belief that rational actions must be uncorrupted by desire and not particular to one's circumstances has given way to a more flexible account of how rationality might underpin autonomy.[102] Morreim prefers the term 'rational competence' which 'concerns our ability to function as rational beings, to make judgments and choices in rational ways'.[103] According to Raz, autonomy includes a minimum rationality which consists of the ability to comprehend the means to realize one's goals and the mental faculties necessary to plan actions.[104] Savulescu and Momeyer argue that being auton-omous requires that a person holds rational beliefs.[105] Lindley terms the

99   Ibid. 297.
100   Ibid. 298.
101   Ibid.
102   See, for example, J. Benson, 'Who is the Autonomous Man?' 58 (1983), *Philosophy* 5; L. Haworth, *Autonomy: An Essay in Philosophical Psychology and Ethics*, New Haven, Conn: Yale University Press (1986); R. Lindley, *Autonomy*, London: Macmillan (1986); R. Rhodes, 'Genetic Links, Family Ties, and Social Bonds: Rights and Responsibilities in the Face of Genetic Knowledge' 23 (1998), *Journal of Medicine and Philosophy* 10; J. Harris and K. Keywood, 'Ignorance, Information and Autonomy' 22 (2001), *Theoretical Medicine* 415. Young does not reject the notion of rationality as an aspect of autonomy out of hand. Indeed, he states: 'Now it is true that many a heteronomous person does behave in a way that fails of rationality. And to that extent rationality seems like a positive require-ment for autonomy'. However, he notes that 'for most of us at some time, acceding to the demands of the rational is liable to restrict our occurent autonomy and perhaps thereby our global autonomy'. R. Young, 'Autonomy and Socialization' 89 (1980), *Mind* 570, 567 and 568 respec-tively. Although MacKenzie and Stoljar do not adopt an individualistic interpretation of autonomy, they seem to reject the idea that rationality is an aspect of autonomy on the basis that rationality is a distinctly male characteristic. C. Mackenzie and N. Stoljar, 'Autonomy Reconfigured' in C. Mac-kenzie and N. Stoljar (eds), *Relational Autonomy: Feminist Perspectives on Autonomy, Agency and the Social Self*, New York: Oxford University Press (2000), 5. See also N. Stoljar, 'Autonomy and the Feminist Intuition' in C. Mackenzie and N. Stoljar (eds), *Relational Autonomy: Feminist Perspectives on Autonomy, Agency and the Social Self*, New York: Oxford University Press (2000), 106, where she argues that historically women have been characterized as emotional rather than rational.
103   H. Morreim, 'Three Concepts of Patient Competence' 4 (1983), *Theoretical Medicine* 231, 234.
104   J. Raz, *The Morality of Freedom*, New York: Oxford University Press (1988), 373.
105   J. Savulescu and R. W. Momeyer, 'Should Informed Consent be Based on Rational Beliefs?' 23 (1997), *Journal of Medical Ethics* 282.

rationality that he views as a proper condition of autonomy as 'active theoretical rationality'. He argues that this requires an individual to take an active role in the truth of her beliefs and the validity of her desires.[106] In a similar vein, Haworth has developed a model of rational autonomy which relies on reasoning capacity which requires that one goes 'far enough in finding reasons of one's preferences without needing to go to heroic lengths of deliberating endlessly'.[107] In the face of such varied accounts of rationality, the question of what the right thing is is likely to depend on the identity of the rationality assessor.

However, rationality is not a subjective notion. Just because I think I am rational does not mean that I am, but, neither is there a clear, shared objective interpretation of rationality. There is vast literature on the theory of rationality, and the conceptions of rationality in this literature are varied.[108] The variation is so much that Goldman argues that the notion of rationality is so disparately employed by different philosophers and social scientists, that it has limited usefulness.[109] This discussion does not aim to provide a comprehensive account of the theory of rationality.[110] Far from it, here the aim is simply to introduce and consider the notions of substantive or value rationality and procedural rationality from the perspective of the functionality of these notions as aspects of the principle of autonomy, which might form the basis of recognizable harm in the tort of negligence.

In his seminal work on rationality, Weber acknowledges a fundamental distinction between substantive or value rationality and instrumental rationality.[111] Brownsword draws the same distinction when arguing for a

---

106 R. Lindley, *Autonomy*, London: Macmillan (1986), 63–70.

107 L. Haworth, *Autonomy: An Essay in Philosophical Psychology and Ethics*, New Haven, Conn: Yale University Press (1986), 39.

108 Lene Koch argues that we operate with a number of cultures, a number of world views and a number of corresponding rationalities. L. Koch, 'IVF – An Irrational Choice?' 3 (1990), *Journal of International Feminist Analysis* 235.

109 A. I. Goldman, *Epistemology and Cognition*, Cambridge, MA: Harvard University Press (1986) 27.

110 There are many foundational works on the concept. A few of these are: M. Weber, *Rationality and Modernity*, S. Whimster and S. Lash (eds), London: Routledge (2008); R. Nozick, *The Nature of Rationality*, New Jersey: Princeton University Press (1993); K. S. Cook and M. Levi (eds), *The Limits of Rationality*, Chicago: University of Chicago Press (1990).

111 M. Weber, *The Theory of Social and Economic Organization*, New York: The Free Press (1964); Nozick draws a similar distinction: R. Nozick, *The Nature of Rationality*, New Jersey: Princeton University Press (1994). Within the context of this distinction, Weber further draws a distinction between four types of rationality: Practical rationality, which involves the individual who considers ends, and on some systematic basis decides what is the best means or course of action to pursue in order to achieve these ends; theoretical rationality which requires that abstract concepts or theoretical models form an essential part of logical reasoning; substantive rationality which requires that individuals attempt to make their values and actions consistent; and formal rationality which is a broader form of rationality that characterizes organizations, especially bureaucratic ones.

rational theory of contract law.[112] Whilst the former incorporates a requirement that choices, actions and desires be somehow legitimate, the latter
requires that an individual act in accordance with her desires *whatever* they
may be. The latter does not pour content into choices, actions and desires,
but it requires that the process of achieving them meets certain standards
designed to ensure that a person's true or authentic desires are effective in
action.[113] Let us consider how each of these rationality perspectives might
function as a condition of autonomy.

Substantive/value rationality

Substantive or value rationality is focused on ultimate ends.[114] It ignores the
process by which those ends are achieved; the end is rational if it meets with
some objective notion of what rationality consists in. Thus, the rationality of
a particular choice, action or desire is determined by reference to the essence
of the end which it seeks to achieve. Although reasons are sometimes seen as
the hallmark of rationality,[115] value rationality does not consist in individual
reasoning. Focusing on the end itself, it is irrelevant whether the right
reasons or, indeed, if any reasons, led the individual to arrive at the particular end. If the question of whether an end is rational is to be determined by
reference to the value of the end itself, there must be some objective standards or norms which allow a distinction to be made between rational and
non-rational ends. Indeed, Nickerson argues: 'To say that this behaviour is
rational and another is not implies the existence of standards on which such
a distinction can be based.'[116]

   Without such standards, the question of what is rational is persistently
moot. Thus, empirical explanation of what constitutes rationality depends
on the conception of value adopted to form the content of rationality. In the
context of the theory of contract law, Brownsword notes that a theory of
rationality based on substance creates a legitimacy problem.[117] Despite
arguing that it remains unclear whether moral judgments are a necessary

---

112   R. Brownsword, 'Towards a Rational Theory of Contract Law' in T. Wilhelmsson (ed.), *Perspectives
       of Critical Contract Law*, Aldershot: Dartmouth (1993), 241, 248.
113   In Brownsword's model, instrumental rationality describes a goal-orientated process which aims to
       ensure the efficacy of securing goals. Substantive rationality requires that the substance of a rational
       (law on Brownsword's consideration) should be justified or legitimate. R. Brownsword, 'Towards a
       Rational Theory of Contract Law' in T. Wilhelmsson (ed.), *Perspectives of Critical Contract Law*,
       Aldershot: Dartmouth (1993), 241, 248.
114   Hereafter, the term value will largely be used.
115   R. S. Nickerson, *Aspects of Rationality: Reflections on What it Means to be Rational and Whether We Are*,
       New York: Taylor & Francis (2008), 13.
116   Ibid. 37.
117   R. Brownsword, 'Towards a Rational Theory of Contract Law' in T. Wilhelmsson (ed.), *Perspectives
       of Critical Contract Law*, Aldershot: Dartmouth (1993) 241, 250.

element in rationality judgments,[118] Brownsword argues that in the context of contract law, if substantive rationality is understood in terms of moral objectivism and rationally defensible moral criteria, Gewirth's argument to the PGC can be relied on to pour particular content into the idea of substantive rationality. This assumes that the argument to the PGC is valid and that there is agreement in its interpretation.[119]

Given that the notion of value rationality is being considered here from the perspective of its function as an aspect of the interpretation of autonomy within English negligence law, it might be argued that the English negligence system holds an intrinsic notion of value, which could be relied on to pour content into the conception of rationality, which might form an aspect of autonomy. English negligence law is deeply imbued with an objective notion of value based on the perception of the position of the ordinary or reasonable person. This standard could arguably be used as a measure of rationality in the same way that it is used as a measure of reasonableness. Those who adopt an individualistic account of autonomy might argue that the benchmark of the ordinary person is inimical to autonomy, because autonomy is 'essentially a personal concept, which is supposed to take account of the individual's own interests and concerns – not those of the homogenised, fictional character'.[120] Nevertheless, 'it is very much the common pattern of the law to establish and develop general concepts such as the "reasonable man" (or the "prudent patient") in order to adjudicate on disputes'.[121] Whilst this common legal standard might be inimical to individualistic accounts of autonomy, it is arguably complementary to moral accounts of autonomy based on rationality. This is not to suggest that rationality and reasonableness are the same thing.[122] The courts assume that the ordinary person is reasonable when setting the standard for breach of duty in negligence,[123] or questioning whether the law ought to recognize a particular kind of loss,[124] here they would be making a similar assumption that the ordinary person is rational.[125]

---

118 Ibid. 268.

119 Which Brownsword readily acknowledges.

120 S. McLean, *Autonomy, Consent and the Law,* London: Routledge Cavendish (2010), 83.

121 Ibid. 83.

122 A. Gewirth, 'The Rationality of Reasonableness' 57 (1983), *Synthese* 225. However, Gewrith argues that both are based on the notion of reason.

123 The notion of the reasonable man is the objective standard of care in negligence which allows evaluation of whether the defendant's actions were of a standard so as not to be regarded as careless. *Blyth v Birmingham Waterworks Co* (1856) 11 Ex. 781; *Glasgow Corp v Muir* [1943] AC 448.

124 *McFarlane v Tayside Health Board* [2000] 2 AC 59, Lord Steyn, 82; *White v Chief Constable of South Yorkshire Police* [1999] 2 AC 455, Lord Hoffmann, 510–511.

125 For example, it is foundational to Kant's theory that all humans are rational beings: I. Kant (1785), *The Groundwork for the Metaphysics of Morals,* translated by M. J. Gregor, Cambridge: Cambridge University Press (1998), 428.

Although the notion of the reasonable man[126] is predominantly relevant to the question of whether the defendant's actions were careless, the notion of the values of the ordinary or reasonable man pervades the whole of negligence law. This notion of value is often applied to the question of whether the interest which is interfered with is of sufficient value to warrant legal recognition.[127] There may be better ways for the courts to determine whether a claim seeking to rely on a novel interest has value. However, this discussion does not claim that the court's mechanism of relying on its perceived view of the reasonable person is the way in which the courts ought to deem what is valuable and, therefore, deserves legal recognition. It claims that it *is* a way which the courts determine value.[128]

There have been many cogent criticisms of the assessment of the value of novel claims by reference to the standard of reasonable or ordinary people. Unlike the application of the reasonable or ordinary man standard in the context of breach of duty, the application with respect to the value of the interest to be protected by the claim itself is not referred to the particular body of ordinary or reasonable people, whose values the law purports to reflect.[129] It relies instead on the courts' perception of the view of the reasonable man which might not be a reflection of the actual position, if, in fact, there is a societal consensus as to what is reasonable.[130]

Value pluralists believe that different values are not somehow reducible to one supervalue. From this pluralist perspective, many values can be

---

126 The reasonable man and the rational man might not be identical concepts but for the purposes of this analysis they do not need to be. The point is that the court might adopt what it perceives to be the views of the ordinary person to form the basis of the values which it takes as pouring content into the notion of value rationality.

127 For particularly strong judicial assertions tying the value of the claimant's action to the perceived view of the popular majority, see, for example, the decisions of the House of Lords in *McFarlane v Tayside Health Board* [2000] 2 AC 59 where Lord Steyn thought it was relevant to ask commuters on the Underground the following question: 'Should the parents of an unwanted but healthy child be able to sue the doctor or hospital for compensation equivalent to the cost of bringing up the child for the years of his or her minority, i.e. until about 18 years?' He was firmly of the view that an overwhelming number of ordinary men and women would answer the question with an emphatic 'No'. Lord Steyn, 82. In *White v Chief Constable of South Yorkshire Police* [1999] 2 AC 455, Lord Hoffmann refused to extend the law because he thought that such an extension would be unacceptable to the ordinary person. Lord Hoffmann, 510. In *Re MB (Medical Treatment)* [1997] 2 FLR 426, Butler-Sloss LJ (as she then was), 437, said that irrationality means a decision which is so outrageous in its defiance of logic or of accepted moral standards that no sensible person who had applied his mind to the question to be decided could have arrived at it.

128 It might be argued that it is sensible for a legal system to be imbued with the values which are shared by the majority of its subjects. The difficulty lies in determining whether people do, in fact, have such shared values and whether judges can accurately determine what these are.

129 I.e. other people with the particular specialism of the defendant.

130 For reliance on societal consensus as a means of determining what should amount to legally recognizable damage, see, for example, *Turpin v Sortini* (1982) 643 P 2d 954, Kaus J, 963.

equally right and, therefore, conflict with one another.[131] In these circumstances, values are incommensurable on the basis that they cannot be put into an objective hierarchy of values.[132] Following on from this, pluralists believe that a reasoned choice between conflicting values cannot be made.[133] Berlin argues that the ideal of harmony is unobtainable and, in fact, incoherent because securing or protecting one value necessarily involves abandoning or compromising another.[134] The question which follows is whether there is conflict in what the truth is, or conflict between people as to what the truth is. For Berlin, the conflict was in what the truth actually is. Ronald Dworkin refutes value pluralism and argues that one ultimate value can be discovered in the face of an apparent conflict between values. Indeed, according to Dworkin, everyday people, and for that matter everyday judges, find their way and choose between options and values that were supposed to be incommensurable. Dworkin believes that through reliance on other rules and principles, it is possible to redefine and harmonize the meanings of the values involved in order to resolve the conflict between the seemingly incommensurable values.[135] From this perspective, in any particular circumstance, there is a definition of the value in question that we should accept leads us to discover the correct hierarchy of values. Thus, there is a true or ultimate value that emerges which depends on our having accepted the correct definition of the relevant values in the first place.

Although the courts might in abstract recognize the existence of value pluralism,[136] the law is a practical animal. If it is not practical, it has utterly failed,[137] whatever or however many values it is prepared to recognize and endorse. Thus, where there are values that conflict, the law must have a way of defining those values so as to find an ultimate value. This sociological method of giving content to the notion of value rationality will be problematic for those who do not believe in a shared notion of the common

---

131 T. Parsons, *The Structure of Social Action*, New York: The Free Press (1968).

132 J. Raz, *The Morality of Freedom*, New York: Oxford University Press (1988).

133 I. Berlin, 'Two Concepts of Liberty' in I. Berlin, *Four Essays on Liberty*, London: Oxford University Press (1969); D. Wiggins, 'Incommensurability: Four Proposals' in R. Chang (ed.), *Incommensurability, Incomparability and Practical Reason*, Cambridge, MA: Harvard University Press (1997); J. Kekes, *The Morality of Pluralism*, New Jersey: Princeton University Press (1993).

134 I. Berlin, 'Two Concepts of Liberty' in I. Berlin, *Four Essays on Liberty*, London: Oxford University Press (1969).

135 R. Dworkin, *Justice in Robes*, Cambridge, MA: Harvard University Press (2006), 111–113.

136 The courts regularly have to deal with competing values where there is no clear answer to the question of which value should be recognized as the ultimate value. See *Rees v Darlington Memorial Hospital NHS Trust* [2004] 1 AC 309, Lord Steyn, 324. In particular where Lord Steyn said 'The issue in *McFarlane* was a profoundly controversial one. Ultimately, there was a choice to be made between eminently reasonable competing arguments'.

137 C. Foster, *Choosing Life, Choosing Death: The Tyranny of Autonomy in Medical Ethics and Law*, Oxford: Hart (2009), 13.

good.[138] However, some commentators believe that socialization leads to the holding of collective social values.[139] Parsons, in particular, argues that the notion of shared values accounts for the regularity and determinacy that is characteristic of everyday social interactions.[140] This notion of a majority value on behalf of the subjects of a particular legal system provides the courts with a method of defining values and harmonizing them so as to discover the ultimate value. If one adopts the position that shared socialization leads to shared values, within a particular society, there is likely to be agreement about whether a particular choice, action or desire is valuable and, therefore, rational. The accepted ultimate value on a given matter might then provide a basis upon which the courts could pour content into the notion of value rationality.[141]

Although value rationality does not consist of individual reasoning,[142] it might be argued that by focusing on the reasons for an individual's desire we can give an alternative account of value rationality, which focuses on reasons as indicators of value as opposed to ends in themselves as indicators of value. End point value rationality focuses on objective values rather than the particular values of the individual. The requirement that the individual demonstrates valuable reasons for her ends would mean that the end in itself need not accord with some notion of rational value, but the reasoning of the particular individual would be relied on to indicate value. This approach ties the notion of value to the particular individual and her circumstances, allowing a more subjective concept of value to be adopted than that which might apply to the question of whether a particular end is rational per se. This approach might have particular application where the end itself appears odd and therefore might be perceived by many as irrational.[143] Where this is the case, a particular end's value might be determined through the individual's reasons for that end.

Morgan and Veitch have discussed how, despite express statements to the contrary, in the context of medical law, the courts do actually require that patients demonstrate, to the satisfaction of the court, that they have reasonable reasons for their decision before they will deem that decision rational.[144]

---

138 K. Ansell-Pearson, 'Nietzsche on Autonomy and Morality: The Challenge to Political Theory' 39 (1991), *Political Studies* 270, 270.

139 See T. Parsons, *The Structure of Social Action*, New York: The Free Press (1968).

140 See, for example, T. Parsons, *The Structure of Social Action*. New York: The Free Press (1968).

141 This approach ignores the question of the difficulty of ascertaining the perceived majority view. There may be significant difficulties associated with this but this issue is beyond the scope of this enquiry.

142 See discussion above.

143 As is essentially the case in many high profile medical law cases. See, for example, *Re B (Consent to Treatment: Capacity)* [2002] 1 FLR 1090; *Re MB (Medical Treatment)* [1997] 2 FLR 426; *St George's Healthcare NHS Trust v S; R v Collins Ex p. S (No. 2)* [1999] Fam 26; *Re E (A Minor) (Wardship: Medical Treatment)* [1993] 1 FLR 386.

144 D. Morgan and K. Veitch, 'Being Ms B: B, Autonomy and the Nature of Legal Regulation', *Sydney Law Review* (2004), 107.

They argue that, despite the fact that the English courts have repeatedly said that in the context of medical treatment the patients' right to choose or to refuse treatment is not limited to decisions which others might regard as sensible, it exists notwithstanding that the reasons for making the choice are rational, irrational, unknown or even non-existent.[145] The assessment of mental capacity and the nature of patients' decisions merge, so individuals must, in fact, explain the reasons for their decisions.[146] They focus on the case of *Re B (Consent to Treatment: Capacity)*[147] to argue that in order to justify her choice, Ms B had to speak of her decision and the reasons behind it. Effectively, in the absence of a decision which would objectively be regarded by others as rational in and of itself,[148] Ms B had to prove to others, in particular the court, that the reasons for her decision were rational. Ms B had to talk of her situation and the level of suffering she was experiencing and convince others that, within the context of her suffering, reasons existed upon which a rational decision to pursue the end of her life could be formed. In essence, the fact that she had reasonable reasons for an apparently irrational decision overrode initial doubts about the rationality of her apparently odd decision, and led the court to respect her decision as autonomous on the basis that it was rational because of the reasons behind it. This would allow the courts some control over what autonomy consists in, without requiring that the end itself is rational.[149]

A procedural account of autonomy

Content-neutral theories of autonomy find their core meaning in the idea of being one's own person, imbued with considerations, desires, conditions and characteristics that are not imposed externally on one, but are one's own. The focus is on the person's competent self-direction, free of manipulative and 'external' forces – in a world of 'self-government'.[150] Here the central value is the ability to determine one's way of life for oneself.[151] This individualistic

---

145 See, in particular, *Re T (Adult: refusal of treatment)* [1993] Fam 95, Lord Donaldson, 102; *Sidaway v Board of Governors of the Bethlem Royal Hospital and the Maudsley Hospital and Others* [1985] AC 871, Lord Templeman, 904; *Re MB (Medical Treatment)* [1997] 2 FLR 426, Butler-Sloss LJ, 432.

146 D. Morgan and K. Veitch, 'Being Ms B: B, Autonomy and the Nature of Legal Regulation', *Sydney Law Review* (2004), 107.

147 *Re B (Consent to Treatment: Capacity)* [2002] 1 FLR 1090.

148 Because it was a decision which rejected medical advice and would result in death.

149 For a deeper discussion of autonomy and the subjective character of experience, see K. Atkins, 'Autonomy and the Subjective Character of Experience', 17 (2000), 71, and the discussion of this issue in Chapter 4.

150 J. Christman and J. Anderson (eds), *Autonomy and the Challenges to Liberalism*, Cambridge: Cambridge University Press (2005), 2.

151 G. F. Gaus, 'The Place of Autonomy Within Liberalism' in J. Christman and J. Anderson (eds), *Autonomy and the Challenges to Liberalism*, Cambridge: Cambridge University Press (2005).

content-neutral notion of autonomy is sometimes termed 'liberal autonomy'.[152] It is the kind of autonomy envisioned by philosophers such as Nietzsche, Gerald Dworkin, Berlin and Christman.[153] However, although this interpretation of autonomy does not require that a choice, action or desire in itself meet particular standards, it requires that the autonomous person possess a character whereby she has the will to order her desires.[154] Thus, although ends in themselves do not have to accord with particular notions of value, certain procedural standards are required to ensure that a person's choices, actions and desires represent her true or 'authentic'[155] preferences. Thus, in addition to being free from external constraints such as coercion,[156] this procedural account of autonomy requires that an individual be free of internal constraints, which might prevent her from doing what she really wants to do; namely the inability to critically reflect upon and order her desires and make distinctions which reflect her preferences.[157]

Although this theory is content-neutral, it amounts to more than the basic notion of autonomy whereby it is simply inherent in doing that we do what we want.[158] On this view, autonomy amounts only to procedural independence. There is no need to reflect on our desires at a higher level because if that desire arises in me, free from the undue influence of others, the desire is my own, and in acting upon it, I act autonomously. Although procedural accounts of autonomy do not rely on substantive content, because the procedure itself must meet certain standards, the concept of autonomy does rest on some conditions which are capable of external evaluation. That is some form of rational reflection on the desires that become effective in action, as opposed to indifference to the enterprise of evaluating one's desires.[159]

Procedural accounts of autonomy hold that procedural independence is necessary for autonomy, but that it is not sufficient.[160] Gerald Dworkin presents a particularly developed procedural account of autonomy. According

---

152   Ibid. 272.

153   See above discussion.

154   R. Young, 'Autonomy and Socialization' 89 (1980), *Mind* 570. The ability to reflect on and order one's desires also forms the core of Frankfurt's and Dworkin's interpretations of autonomy.

155   To coin Dworkin's term.

156   Gerald Dworkin terms this aspect of autonomy 'procedural independence': G. Dworkin, 'Autonomy and Behaviour Control' 6 (1976), *Hastings Center Report* 23, 25.

157   Gerald Dworkin terms this aspect of autonomy 'authenticity': G. Dworkin, 'Autonomy and Behaviour Control' 6 (1976), *Hastings Center Report* 23, 24.

158   See, for example, J. P. Plamenatz, *Consent, Freedom and Political Obligation*, London: Oxford University Press (1938), 110; T. V. Daveney, 'Wanting' 11 (1961), *Philosophical Quarterly* 139.

159   H. Frankfurt, 'The Freedom of the Will and the Concept of a Person' in *The Inner Citadel*, J. Christman (ed.), Oxford: Oxford University Press (1989), 69.

160   G. Dworkin, 'Autonomy and Behaviour Control' 6 (1976), *Hastings Center Report* 23, 25.

to Dworkin, the autonomous person is one who does her own thing, which requires characterizations of what it is for a motivation to be *hers* and what it is for it to be her *own*. Dworkin calls the former authenticity and the latter independence.[161] The discussion below focuses on the ability of authenticity to be conceived as an element of procedural rationality because of its emphasis on consistency, which is a key element of rationality.[162]

Philosophers recognize that people may have desires which could at times conflict with one another.[163] Dworkin distinguishes between first order considerations and second order judgments and argues that autonomy cannot be located on the level of the first order considerations. He believes that a crucial feature of a person is their ability to reflect on, and adopt, attitudes towards first order desires.[164] Thus, one might have a particular desire, but also have a desire not to have that desire.[165] Dworkin argues that it is a necessary condition for being autonomous that, at the second order, the person raises and questions whether she will identify with, or reject, the reasons upon which her first order actions are based.[166] On this analysis, it follows that if she rejects the reasoning for her first order desire at the second level, she is not acting autonomously. Frankfurt takes a similar approach, which argues that besides wanting and choosing and being moved to do this or that, men may also want to have (or not to have) certain desires and motives. They are capable of wanting to be different in their preferences and purposes from what they actually are.[167] For both Dworkin and Frankfurt it is central to the principle of autonomy that people are able to critically reflect at some higher level on their first order desires and form preferences concerning their desires. A person then acts autonomously when she is moved to act in accordance with what she discovers she prefers through this reflection. Dworkin and Frankfurt argue that this kind of critical reflection

---

161 Ibid. 25.

162 See, for example, H. Gensler, 'Introduction to Logic', London: Routledge (2002): M. A. Kaplan, 'Means/Ends Rationality', 87 (1976) *Ethics* 61; A. Gewrith, 'The Rationality of Reasonableness' 57 (1983), *Sythese* 225.

163 G. Dworkin, *The Theory and Practice of Autonomy*, Cambridge: Cambridge University Press (1988); H. Frankfurt, *The Importance of What We Care About*, Cambridge: Cambridge University Press (1988); and R. Noggle, 'Autonomy Value and Conditioned Desire' 32 (1995), *American Philosophical Quarterly* 57.

164 G. Dworkin, *The Theory and Practice of Autonomy*, Cambridge: Cambridge University Press (1988), 15; G. Dworkin, 'Acting Freely' 4 (1970), *Nous*, 367.

165 G. Dworkin, *The Theory and Practice of Autonomy*, Cambridge: Cambridge University Press (1988), 15. Dworkin gives the example of the unwilling smoker.

166 Ibid.

167 H. Frankfurt, *The Importance of What We Care About*, Cambridge: Cambridge University Press (1988), 12. It is worth noting that many commentators criticize this approach on the basis that it creates the problem of infinite regress. However the purpose here is not to provide a comprehensive examination of hierarchical accounts of autonomy. Thus I do not consider these critiques.

denotes rationality.[168] It is suggested here that because the hierarchical theory rests on a notion of consistency between the desires and the motivations which move one to action, it reflects a central aspect of rationality; that of consistency.

The dominant rationality concept of the theory of individual rational decision-making is the optimization of expected utility.[169] A choice is rational if the reason for it is that it amounts to acting optimally in pursuit of one's goal.[170] A less onerous interpretation of rationality as consistency is described here, which requires that an individual's motivations are consistent with her authentic desires, as opposed to the optimal means of achieving those desires.[171]

For the most part, people value consistency between goals and goal attainment. As Koehler notes: 'A great deal of work in social psychology has demonstrated that people wish to be consistent in their attitudes and behaviour and are willing to alter one or the other to maintain a consistent state.'[172] People's motivations which are effective in action will usually be consistent with their higher order desires. However, it is not inconceivable that people will make choices that are inconsistent with their higher order desires for reasons of fear, ignorance or addiction, for example.

Imagine a person wants to lead a drug free life and obtain the benefits which go along with that life.[173] Imagine, however that she chooses to take drugs because she wants to be well regarded among her peers or because of addiction. Her preference is to live a drug free life, taking drugs is inconsistent with this preference, thus her choice to take drugs is, from the vantage point of rationality as consistency, irrational. One might argue that she has different, more immediate desires; to be well regarded amongst her friends

---

168   H. Frankfurt, 'Freedom of the Will and the Concept of a Person' in J. Christman, (ed.), *The Inner Citadel: Essays on Individual Autonomy*, New York: Oxford University Press (1989), 63, 68; G. Dworkin, 'Autonomy and Behavior Control' 6 (1976), *Hastings Center Report* 27; G. Dworkin, 'Free Agency', LXXII (1975), *The Journal of Philosophy* 205, 207–208.

169   B. Skyrms, *The Dynamics of Rational Deliberation*, Cambridge, MA: Harvard University Press (1990), vii.

170   For discussions of rational choice theory based on optimization see, for example, G. S, Becker, *The Economic Approach to Human Behaviour*, Chicago: University of Chicago Press (1976); J. Elster, 'Rationality and the Emotions' 106 (1996), *The Economic Journal*, 1386, 1391; K. S. Cook and M. Levi (eds), *The Limits of Rationality*, Chicago: University of Chicago Press (1990), 20; D. Schmidtz, 'Rationality within Reason' 89 (1992), *The Journal of Philosophy* 445–466; D. Parfit, *Reasons and Persons*, Oxford: Clarendon Press (1984), 94; *The Psychology of Human Thought*, R. J. Sternberg and E. E. Smith (eds), Cambridge: Cambridge University Press (1988), 155; J. Kozielecki in *Psychological Decision Theory*, Springer (1982).

171   V. C. Walsh, *Rationality, Allocation and Reproduction*, Oxford: Oxford University Press (1996). In particular, 81.

172   D. J. Koehler, 'Explanation, Imagination and Confidence in Judgment' 110 (1991), *Psychological Bulletin* 499, 508.

173   As opposed to a life on drugs.

or to get high. Thus, there is a conflict between the desire which moves the individual to action and what she mostly wants. She wants to want to lead a drug free life. She does not want to want to get high or want to want to be motivated by peer pressure. She cannot simultaneously want to want to remain drug free and want to want to get high.

For Frankfurt and Dworkin, the desire to do something (here take the drug) would be a 'first order desire', but a unique feature of humans is, Frankfurt argues, that they are capable of forming desires of the second order; that is wanting to have a certain desire or wanting a desire to be one's will.[174] From this vantage point, Frankfurt concludes that the unwilling addict identifies himself, through the formation of second order volition, with one rather than the other of his first order desires (here the desire to take the drug or the desire not to take the drug). In doing this, the individual makes one of his first order desires more truly his own and withdraws from the other.[175] For Frankfurt, the distinction between first and second order desires is used to establish when a person is acting of her own free will. In virtue of this identification and withdrawal established through the formation of second order volition, Frankfurt argues that the unwilling addict can claim that the force moving him to take the drug is a force other than his own, and that it is not of his own free will, but rather against his will that this force moves him to take it.[176]

According to Dworkin, it is a characteristic of persons that they are able to reflect on their desires and in doing so form preferences concerning these. He gives the following examples of how desires might be preferentially ranked: a person may not only desire to smoke, she can also desire that she desires to smoke. However, she may on the other hand not want to have the desire to smoke.[177] In each case what she wants to want is what the individual recognizes as her true self, and is the wish that she wants to see carried out.[178] That is, although the individual has the desire of the first order, she may not want that desire to be effective in the form of action or choice.

Like the notion of value rationality as an aspect of autonomy, this procedural interpretation of rational autonomy provides some analytical purchase on autonomy over and above that which might be possible on the basic account of autonomy, which is devoid of substantive or procedural conditions. For this reason, it might be attractive to a court seeking to recognize the interest in autonomy in a way which is comprehensive and capable

174 H. Frankfurt, *The Importance of What We Care About*, Cambridge: Cambridge University Press (1988), 16.

175 Ibid. 18.

176 Ibid.

177 G. Dworkin, 'Autonomy and Behaviour Control' 6 (1976), *Hastings Center Report* 23, 24; G. Dworkin, *The Theory and Practice of Autonomy*, Cambridge: Cambridge University Press (1988), 15.

178 G. Dworkin, 'Autonomy and Behaviour Control' 6 (1976), *Hastings Center Report* 23, 24.

of being contained. Moreover, if the courts were to adopt this interpretation as the basis for the recognition of the interest in personal autonomy, it would entail the protection of choices which relate to deeply held authentic desires that are the result of critical reflection, as opposed to wanton desires that would be recognized as autonomous on the basic liberal account of autonomy outlined above. According to Frankfurt, the essential characteristic of a wanton is that his desires move him to do certain things, without it being true of him that he wants to be moved by those desires or that he prefers to be moved by other desires.[179] Authentic choices will, however, reflect what the individual deeply values as opposed to what she, perhaps thoughtlessly, desires.

If the courts were to adopt a conception of autonomy that required people to demonstrate consistency between their ultimate desires and actions, it may provide a means of preventing claims which are unmeritorious on the basis that the possibility of compensation has led the claimant to act on a desire to receive compensation that does not reflect her own deeply held values. Tangible tortious losses, corporeal or otherwise,[180] are usually easily ascertainable by the outside world. In some of the claims considered in this book the potential interference with autonomy occurs prior to any explicit exercise of autonomy by the claimant. The question of whether autonomy is, in fact, interfered with depends on what the claimant's desires actually are and whether the defendant's actions did, by chance, interfere with those desires. Imbuing the principle of autonomy with a notion of procedural rationality based on authenticity could protect self-government by facilitating the protection of the individual's authentic desires, and recognizing those desires as autonomous only if the individual can demonstrate that she identifies with them as discussed here.

Intrinsic or instrumental value?

Assuming that English negligence law accepts that the interest in autonomy ought to be recognized as actionable damage, the next step will be for the courts to decide what it is that is so valuable about autonomy which deserves legal protection. Does the value of autonomy derive from the things it makes possible? Or does autonomy have value independent of the effects it produces; that is, is autonomy worth experiencing for its own sake? This discussion should be distinguished from the discussion of substantive autonomy in the previous section. Here the concern is not whether a person's ends reflect particular objective values, it is whether autonomy per se is valuable as an end in itself, or whether it is only valuable as a means to achieving one's desired end, whatever that end might be.

---

179   H. Frankfurt, 'Freedom of the Will and the Concept of a Person' in *The Inner Citadel: Essays on Individual Autonomy*, J. Christman, (ed.), New York: Oxford University Press (1989), 67.
180   Largely financial.

An intrinsic good is a good that is not a means to some further good. If one adopts the intrinsic position, a failure to respect one's desires or present the full range of options would interfere with that person's autonomy even if that failure has no further ill effects on a relevant end which the individual seeks to achieve. From an instrumental perspective, the value of autonomy lies in the objectives which the exercise of autonomy seeks to secure. If by some lucky happenstance the person's ultimate plan is not frustrated by the interference, from an instrumental perspective, her autonomy is not interfered with. However, if the value of autonomy is purely instrumental, then it might be argued that autonomy is largely a principle which can be relied on as a vehicle for the recognition of other values.[181] The true value of autonomy, which cannot easily be explained on the basis of other values, is arguably intrinsically situated.

Intuitively, where autonomy pertains to choice, it is thought to involve choice from a range of options. That is, you are more autonomous if you can choose from all ten available options in a particular situation, than if somebody reduces the number of options available to you in that situation to just one. However, Hurka asks is the individual's autonomy interfered with if all the options but one are taken away from the individual, but the one option that is left open to her is the option that she would have ranked most highly, and therefore chosen, amongst the ten?[182] The instrumental value of autonomy has not been interfered with here because the individual is still able to do what she wanted to do. In terms of making her desires effective in action, there are no ill effects from the reduction in options. However, it might be argued that the individual has lost something of value in the loss of the opportunity to make a choice between options per se.

The issue of disclosure of medical risk elucidates this distinction within the context of tort law. Assume that Adam suffers ongoing back pain. He consults a physician who suggests that an operation will alleviate this pain. In obtaining consent for the operation, the physician negligently fails to tell Adam about the significant risk that the operation will cause some level of paralysis. Further, assume that Adam would have chosen to continue taking painkillers rather than subject himself to the risk of paralysis had he known of the risk. At this point, we might recognize that there has been an interference with Adam's autonomy on the basis that possession of the information relevant to a decision is a core feature of autonomy. However, if we analyse the question of whether Adam's autonomy has been interfered with from the post-operative perspective, then the outcome of the operation may be relevant to the question of what it is of value that he has lost. Suppose the risk eventuates; we might say Adam would never have opted to have the surgery and, therefore, would not have suffered the ill effect of paralysis if he had been given full information; thus he has lost

---

181 Such as freedom from physical or mental harm, or the loss of opportunities or chances.
182 T. Hurka, 'Why Value Autonomy' 13 (1987), *Social Theory and Practice* 361, 362.

something of value in terms of what the exercise of his autonomy would have made possible. But what if the risk does not eventuate? Here, Adam's uninformed decision to have the operation has had no ill effects but this does not mean that he has not lost something of value. The core issue here is not whether interference with autonomy is a *wrong* or a *harm*,[183] the issue is that the question of where the value of autonomy lies is crucial in determining what the harm is. If the true value of autonomy is intrinsic, the individual has been harmed as soon as her choice is not respected. Assume that a health professional acts carelessly or recklessly[184] with regard to the individual's fundamentally important decision, which she has entrusted to the care of that health professional. When the individual discovers this she feels that the lack of respect for her carefully considered decision is detrimental to her self-respect and self-esteem. Loss is central to this account. It is not that the lack of respect for the decision only constitutes the wrong and the loss is dependent on the frustration of what the choice made possible.

Where choice is diminished and full information is not disclosed, the decision-maker may be no worse in terms of what choice was made possible in not being given the full range of choices. That is, things have not turned out differently to how she would have determined them herself. If the harm is only recognized in the frustration of the ultimate end result, autonomy only has instrumental value. However, this fails to capture the view that there is a value connected with being self-determining which is not a matter of bringing about results.[185] If a life without free choice is poorer for that very reason, then autonomy is intrinsically good.[186] Hurka adeptly demonstrates the intrinsic value of autonomy in the context of diminishing choice. He argues that the ideal of agency is one of causal efficacy, of making a causal impact on the world and determining facts about it.[187] When one chooses among options, she more fully realizes this ideal, than one for whom only the favoured action is available. That is because choosing among options has two effects; realising some options and blocking others and this results in a larger efficacy than someone whose only effect is the first.[188] This argument applies to the intrinsic value of choosing from

---

183   That is, it is not only a wrong where the intrinsic value is interfered with but a wrong and a harm where the instrumental value is interfered with.

184   The law of negligence does not deal in degrees of carelessness, so either of these is as culpable as the other.

185   G. Dworkin, *The Theory and Practice of Autonomy*, Cambridge: Cambridge University Press (1988), 112.

186   See, for example, T. Hurka, 'Why Value Autonomy?' 13 (1987), *Social Theory and Practice*; C. H. Wellman, 'The Paradox of Group Autonomy' 20 (2003), *Social Philosophy and Policy* 265, 270; R. Young, 'The Value of Autonomy' 32 (1982), *Philosophical Quarterly* 35; G. Dworkin, *The Theory and Practice of Autonomy*, Cambridge: Cambridge University Press (1988), 6 in particular.

187   T. Hurka, 'Why Value Autonomy?' 13 (1987), *Social Theory and Practice* 361, 366.

188   Ibid.

numerous options. But the same point applies where a choice is motivated by lack of, or erroneous, information. Here, the issue is not that the person's choice would have been different, but the choice is to some extent controlled by an external source; the wrong or lacking information.[189] In this situation, the individual may similarly feel that the causal responsibility for the realization of the outcome does not rest with her.

It seems that the essence of the intrinsic value of autonomy is that being recognized as a person who is capable of being self-determining is somehow central to our self-respect. Our sense of self-respect is likely to be influenced by the respect that others afford to our ability. Being allowed by others to take responsibility for what we are able to choose is an important factor in respecting oneself. Where a choice is influenced by another's careless failure to provide the relevant information, or where there is a wanton disregard for an individual's desires, the individual's sense of responsibility for self-determination will be diminished. Failing to recognize the intrinsic value of autonomy is detrimental to a person's self-respect and self-esteem.[190]

However, despite the fact that a theoretical argument can be made for recognizing the intrinsic value of autonomy, if the interest in autonomy is recognized as the basis for an action in negligence, the courts may wish to limit the scope of the action by only recognizing its instrumental value. The cue for this potential recognition of the interest in autonomy as a basis for an action in negligence is taken from the House of Lords itself.[191] Building on the importance of autonomy in the field of informed consent,[192] the House of

---

189 That is, in the event of having the information, the person would have made the same decision but for different reasons.

190 See R. Young, 'The Value of Autonomy' 32 (1982), *Philosophical Quarterly* 35, 39, for the view that autonomy is intrinsically valuable because it is connected with personal dignity and self-esteem.

191 *Rees v Darlington Memorial Hospital NHS Trust* [2004] 1 AC 309 and *Chester v Afshar* [2005] 1 AC 134.

192 See, for example, *Re T (Adult: refusal of medical treatment)* (1993) Fam 95, Lord Donaldson, 102: '[t]his right of choice is not limited to decisions which others might regard as sensible. It exists notwithstanding that the reasons for making the choice are rational, irrational, unknown or even non-existent'. This commitment to personal autonomy has been reiterated on a number of occasions. See, for example, *Re B (Consent to Treatment: Capacity)* [2002] 1 FLR 1090, Dame Butler-Sloss, 1095, and *Re MB (Medical Treatment)* [1997] 2 FLR 426, Butler-Sloss LJ, 432. For academic support for the importance of autonomy in relation to other principles see, for example, R. Gillon, 'Ethics Needs Principles – Four can Encompass the Rest – and Respect for Autonomy Should be "First Among Equals"' 29 (2003), *Journal of Medical Ethics* 310. Some have criticized 'the extraordinary hegemony of autonomy' in medical law. See, C. Foster, *Choosing Life, Choosing Death: The Tyranny of Autonomy in Medical Ethics and Law*, Oxford: Hart (2009); O. O'Neil, *Autonomy and Trust in Bioethics*, Cambridge: Cambridge University Press (2002); G. Laurie, *Genetic Privacy; A Challenge to Medico-Legal Norms*, Cambridge: Cambridge University Press (2002); G. M. Stirrat and R. Gill, 'Autonomy in Medical Ethics After O'Neill' 31 (2005), *Journal of Medical Ethics*, 127; and D. Callahan, 'Individual Good and Common Good: A Communitarian Approach to Bioethics' 46 (2003), *Perspectives in Biology and Medicine*, 496.

Lords has indicated that autonomy is a fundamental principle,[193] careless interference with which might be protected by the law. It might be argued that if the English legal system considers autonomy to be such an important value, it should recognize its whole value, rather than simply treating the principle as a hook on which to hang other values. The approach suggested in this chapter whereby the principle of autonomy is imbued with an aspect of rationality limits what autonomy itself actually consists in, thereby providing analytical purchase on the principle and allowing the courts to manage liability in a way which does not refuse to recognize part of the value of autonomy in and of itself. That is, it is the principle itself that is limited through the theory of rational autonomy, so that some choices, decisions or actions are simply deemed not to be autonomous, as opposed to a limitation which is imposed despite the fact that a choice, action or desire is recognized as autonomous based on what is thought to be valuable about autonomy. It might be argued that if the courts are to provide meaningful protection for the interest in autonomy, they should not confine its value to that which it makes possible because this undermines the value of experiencing choice and the loss which occurs when choice is not experienced because it is carelessly disrespected.[194]

Conclusion

Autonomy is an important biomedical principle which enjoys legal protection in the context of consent to medical treatment. More recently the House of Lords has recognized careless interference with the concept, paving the way for the argument that interference with autonomy sometimes represents harm in English negligence law. This book discusses four hypothetical genomic claims from the perspective of English negligence law, as if it were explicitly imbued with an interest in autonomy. This chapter considers conceptions of autonomy which might be relied on to form the basis of the principle of autonomy as damage in negligence.

As autonomy is a disparate principle which is subject to many interpretations, this chapter has considered those theories whereby autonomy is conditional upon rationality as substantively or procedurally construed. This conditional approach will allow the judiciary some definitional and conceptual control over the legal development of the principle, which might make them more inclined to recognize that the interference with autonomy is a valid basis for legal harm. Following this, the chapter focuses on what is valuable about autonomy. It argues that the true value of

---

193   *Rees v Darlington Memorial Hospital NHS Trust* [2004] 1 AC 309, Lord Millett, in particular 349, and *Chester v Afshar* [2005] 1 AC 134, Lord Steyn, 146 and Lord Hope, 152.

194   A failure to recognize a patient's choice to have treatment removed, which would lead to her death, would presumably still amount to a harm if she died in any event from some other cause.

autonomy is intrinsic, but concedes that any recognition of autonomy on the courts' behalf is likely to be confined to instrumental value. However, if the very principle of autonomy were limited by a rationality condition, it is argued that this would prevent the need for the courts to refuse to recognize part of what is valuable about autonomy in order to limit potential liability.

Chapter 4

Negligence in reproductive genetics

The wrongfully created person and a claim based on the interest in autonomy

Introduction

Advances in modern genetics enable greater control over the genetic makeup of future generations. The Human Fertilisation and Embryology Act 1990 allows potential parents to screen out embryos that possess a genetic disease using preimplantation genetic diagnosis (PGD).[1] Embryos can be tested and deselected where there is a particular risk that the embryo may have a gene, chromosome or mitochondrion abnormality.[2] The Human Fertilisation and Embryology Authority (the Authority) continues to issue licenses for particular disorders which might be screened out,[3] and it has recently significantly increased the scope of PGD to permit de-selection for a number of genetic disorders which do not manifest until adulthood, and which have a lower degree of penetrence.[4]

This chapter considers the challenges that English negligence law might face if embryo testing to avoid the birth of a child with a particular disorder is performed without due care and attention, leading to the birth of a child with the very disorder that the parents sought to screen out. In these circumstances, it is possible that both the parents and the resulting person might be aggrieved. This chapter considers the challenge which the resulting person might bring.[5]

This chapter follows in two parts: Part I considers how English negligence law might currently react to an action on behalf of the resulting person, which is based on the medical team's negligent failure to test the embryo, from which she has developed, properly for the disorder and screen her out

---

1 As amended by the Human Fertilisation and Embryology Act 2008.
2 Schedule 2, s. 1ZA (1) (b).
3 A central list of disorders which can be screened out can be found on the HFEA website. Available HTTP: www.hfea.gov.uk/pgd-screening.html (accessed 11 May 2010).
4 The Authority licenses screening for the BRCA 1 breast cancer gene, familial adenomatous polyposis coli (FAP) and early onset Alzheimer's.
5 The claims of the parents are considered in the following chapter.

accordingly. The focus is on the only English authority on wrongful life; *McKay v Essex AHA* and the court's perception that the claim concerned neither a harm nor a wrong. The motivation which the principle of the sanctity of life provided for this perception is central to the discussion in this part. The second part analyses the claim from the perspective of English negligence law as imbued with an explicit recognition of the interest in autonomy. Here there is an analysis of the gradual devaluation of the concept of sanctity of life in the medical law context. The focus is on how the courts might react to the hypothetical action if it is construed as an interference with the individual's autonomy interests.

Unwanted birth

The current legal position

The provision of in vitro fertilisation (IVF) has not been without mishaps where clinics have transferred the wrong embryo.[6] Assume a couple approaches a licensed clinic for PGD to avoid a genetic disorder. However, there is a negligent failure to screen the embryos. As a result, an embryo which has the disorder which was to be screened out is transferred and carried to term. Assume further that the resulting person is aggrieved and seeks to articulate this grievance as a novel challenge within the tort of negligence?

Where carelessness in selecting an embryo leads to the birth of a person with the very genetic disorder the parents sought to avoid, the medical team cannot be said to have caused the disorder and it was not possible to cure it.[7] Thus, the gist of the claim cannot be the genetic disorder itself. The gist of the claim must be that the individual should never have been born. This

---

6 See, for example, the mix-up that befell Donna Fasano and Deborah Parry-Rogers in a Manhattan fertility clinic. Essentially, one of Deborah's embryos was mistakenly implanted in Donna. More recently in San Francisco, a fertility team mistakenly implanted in Susan Bushweitz the embryo of one of its other patients. Last year a UK couple's last hopes of having another child were destroyed after a mistake meant their final embryo was implanted in another woman. See H. Carter, 'IVF clinic mix-up destroys couple's last embryo and chance of second child', *The Guardian* (Sunday 14 June 2009). Four adverse events at Leeds Teaching Hospitals regarding the treatment of gametes and embryos led to an independent review. B. Toft, *Independent Review of the Circumstances Surrounding Four Adverse Events that Occurred in the Reproductive Medicine Units at the Leeds Teaching Hospitals NHS Trust,* West Yorkshire, Department of Health (2004). The review concluded that the events were caused by a mixture of inadvertent human error and systems failure.
7 It might be possible to cure some medical conditions in embryos through foetal surgery. However, in the face of PGD it is unlikely that the parents will further test the foetus for the condition *in utero*. Furthermore, foetal surgery is only available for a limited number of conditions, many of which are quite different to the conditions which parents might seek to screen out via PGD. The essence of conditions which can be screened out via PGD is that they cannot largely be subsequently cured.

casts the claim as what is known as a wrongful life action.[8] In English law this action is currently considered unmeritorious on the basis that the claimant has suffered no recognizable harm by being born and the defendant has committed no recognizable wrong in failing to prevent her birth.

Mary McKay was born disabled because her mother contracted rubella during pregnancy. Mary's mother underwent a rubella test, but due to a lack of due care and attention, she was wrongly informed that she had not been infected and, therefore, need not consider an abortion on the grounds that the child may have the kind of defects which result from exposure to rubella *in utero*. Mary claimed that the defendant owed her a duty of care when she was *in utero* to advise her mother correctly of the rubella infection so that she might have exercised her right to abortion. The rubella infection itself was not the fault of the defendants, nor could the damage have been reversed in any way. Thus, the essence of Mary's argument was that her life represented damage which should be compensated by the law.

The Court of Appeal[9] held that the claim should be struck out as disclosing no reasonable cause of action.[10] The ability to strike out is used where the court perceives the claim to be devoid of merit. A claim might be unmeritorious in that corrective justice does not point in the claimant's favour. In *McKay* the Court of Appeal rejected the argument that Mary had a claim to corrective justice because she had not suffered any legally recognized damage by being born,[11] and even if life itself could be recognized harm, no duty could be owed by the doctor to the child to encourage its termination because not doing so could not be deemed to be a legal wrong.[12] With neither a wrong nor a harm, there was nothing to correct.

The hypothetical example concerns negligence in PGD, rather than in failing to advise abortion. Nevertheless, the essence of the child's perceived

---

8 The term 'wrongful life' has attracted significant criticism. J. K. Mason, *The Troubled Pregnancy*, Cambridge: Cambridge University Press (2007), 7.

9 Griffiths LJ dissenting.

10 *McKay v Essex AHA* [1982] QB 1166, Ackner LJ, 1187; Stephenson LJ, 1171.

11 Ibid. Ackner LJ, 1189; Stephenson LJ, 1184.

12 Ibid. Stephenson LJ, 1178. According to Griffiths LJ, the predominant reason for refusing to recognize the wrongful life action was the intolerable and insoluble problem that it would create in the assessment of damage. The argument regarding the inassessability of damages had also been relied on to reject wrongful life actions in the US. See, for example, *Gleitman v Cosgrove* 227 A 2d 689 (N.J. 1967), 692. However, in *Curlender v Bio-Science Laboratories* (1980) 165 Cal. Rptr. 447, the Californian Court of Appeal noted that there had been a gradual retreat from the position of accepting impossibility of measuring damages as the sole ground for barring the infant's right of recovery. See also *Harriton v Stephens* [2006] HCA 15 where the High Court of Australia held that it could not be determined whether the child's present life with disabilities represented a loss which should be recognized as constituting actionable damage. See also, K. A. Warner, 'Wrongful life Goes Down Under' 123 (2007), *Law Quarterly Review* 209 and D. Morgan and B. White, 'Everyday Life and The Edges of Existence: Wrongs with No Name or the Wrong Name' 29 (2006), *University of New South Wales Law Journal* 239.

harm is the same. However, the question of whether there is a wrong is quite different. Let us first consider the question of the basis of the wrong in the hypothetical claim as compared to *McKay,* and then analyse the question of whether what was not deemed to be a harm in *McKay* could be in the hypothetical given the changing emphasis in legal values in the ensuing thirty years.

Is there a wrong?

The Court of Appeal interpreted Mary McKay's claim as arguing that the doctor owed her a duty to inform her mother of the desirability of abortion.[13] According to Ackner LJ, such a duty was contrary to public policy because it would 'put doctors under a subconscious pressure to advise abortion in doubtful cases'.[14] Although a doctor might be expected to inform potential parents of the probability that their child will be born with a disability and the nature and severity of that disability, it does not follow that a doctor must advise abortion to meet the required standard of care. Doctors are trained to give medical advice; the question of whether parents can and want to raise a disabled child is not a medical question. From this vantage point, failing to give advice about the desirability of abortion cannot be a culpable activity. The decision whether to terminate a pregnancy should rest with informed parents.[15] However, the health professional could legitimately owe a duty to give full and accurate information about the nature and likelihood of disability and abstract advice about the provision of abortion services.

The hypothetical claim, however, does not concern abortion. The perceived wrong does not relate to advice which is traditionally outside the defendant's professional remit. The wrong concerns a lack of due care and attention within the very professional service that the defendant has agreed to perform. Every health professional owes a duty to perform professional procedures with due care and attention. Thus, unlike *McKay,* this is not a case where there can be no duty because the perceived wrong arises from an action which is not traditionally thought to be within the scope of the defendant's responsibilities. In the hypothetical claim, the parents are aware of the risk that their child will be born with a particular genetic condition and have already made the decision to exercise their right to have embryos screened out.[16] The medical team is entrusted with not selecting embryos with the deleterious gene. The wrong here relates directly to the performance

13 *McKay v Essex AHA* [1982] QB 1166, Stephenson LJ, 1173; Ackner, 1185.
14 Ibid. Ackner LJ, 1187.
15 The choice whether to terminate or not to terminate a pregnancy rests with the mother, but many couples may make this decision together.
16 That is if the particular condition can be lawfully screened out under the Human Fertilisation and Embryology Act 1990.

of the defendant's agreed responsibilities. If the carelessness amounted to a legal wrong, it would not put an impossible pressure on doctors which would influence them to engage in practices which might be regarded as dubious. Thus, despite the finding that no wrong had been done in *McKay*, it will be difficult for the courts to hold that no wrong has been done in the hypothetical. However, although the wrong can be distinguished in the hypothetical, the harm is, on the face of it, the same in both cases. This chapter seeks to develop a different interpretation of the harm in the hypothetical claim which was not as easily established in *McKay*.

Is there a harm?

The view that Mary McKay had not suffered a harm was motivated by the perception that no life, no matter how disabled, could be worse than non-existence. Stephenson LJ and Ackner LJ refused to impose a duty of care on public policy grounds, on the basis that it would make an 'inroad on the sanctity of human life'.[17] The sanctity of life principle holds that the value of human life always exceeds all other values and that all human life is equal. However, it might be argued that the notion of the sanctity of life is in decline both legally and generally. The scientific basis of evolution and creation is increasingly demonstrable,[18] leading an increasing number of people in Britain to describe themselves as atheist, agnostic or a non-believer in God.[19] Those who come from a non-religious perspective might refute the position that all human life is imbued with absolute inherent value.[20] Indeed, the view appears to be growing that life must have certain critical qualities before it can be regarded as being of ultimate value, rather than it being sacred and inviolable in and of itself.[21] Markesinis argues that although the religious purist might not be able to recognize that a disabled

---

17 *McKay v Essex AHA* [1982] QB 1166, Stephenson LJ, 1180; Acker LJ, 1188.

18 For example, the assumption that God is the creator of all life is more difficult in the face of assisted conception and, in particular, the creation, testing, rejection and destruction of embryos on the basis of reproductive choice and research.

19 The principle of the sanctity of life is based on Judaeo-Christian belief that all life has inherent value because it was created by God. However, a 2005 study by Zukerman found that 31–44 per cent of Britons identify as atheist, agnostic or a non-believer in God. P. Zuckerman, 'Atheism: Contemporary Rates and Patterns' in M. Martin (ed.), *The Cambridge Companion to Atheism*, Cambridge: Cambridge University Press (2005); C. Brown, *The Death of Christian Britain: Understanding Secularisation 1800–2000*, London: Routledge (2001); S. Bruce, *God is Dead: The Secularisation of the West*, Oxford: Blackwell (2002); T. Wormsley, 'Kirk Debates Decline in Churchgoers', *The Scotsman* (25 May 2002).

20 Many modern medical law cases are evidence of this. See, in particular *Airedale NHS Trust v Bland* [1993] AC 789; *Re B (Consent to Treatment: Capacity)* [2002] 1 FLR 1090 and the discussion in Part II of this chapter.

21 See the discussion in Part II of this chapter.

life is worse than no life, such abstract dicta about the comparison between life and no life will not appeal to the majority of the population as a means of deciding whether a person has suffered harm.[22] Teff refutes the strict sanctity of life position, noting that although some people will always be resolutely opposed to the recognition of the wrongful life claim, one could formulate a general principle, permitting a claim if, on the child's behalf, the parents would have opted for its non-existence in circumstances where properly informed parents could reasonably have done so.[23]

There was evidence that when the Court of Appeal heard *McKay*, Parliament did not think that life should be regarded as a legal harm. According to s. 1 (1) of the Congenital Disabilities (Civil Liability) Act 1976[24]:

> If a child is born disabled as a result of such an occurrence before its birth … and a person is … answerable to the child in respect of that occurrence, the child's disabilities are to be regarded as damage resulting from the wrongful act of that person and actionable at the suit of the child.

The Act describes an occurrence as one which 'affected the mother during her pregnancy, or affected her or the child in the course of its birth, so that the child is born with disabilities which would not otherwise have been present'.[25] Although the Act could not be relied on directly in *McKay* because the facts occurred before the Act came into force, all three judges in *McKay* thought that it precluded the possibility that a life could be regarded as harm.[26]

The law now permits the screening out of embryos on the basis of genetic, chromosomal or mitochondrial abnormality.[27] Furthermore, the legitimacy of not selecting embryos with particular conditions is carefully considered by the HFEA before it issues any license for embryo screening. Upon receipt of any application:

> A lay summary of the condition will be published on the HFEA website to allow any interested parties to comment on the application. These

22 S. Deakin, A. Johnston and B. Markesinis, *Tort Law*, fifth edition, Oxford: Oxford University Press (2003), 308.

23 H. Teff, 'The Action for 'Wrongful Life' in England and the United States' 34 (1985), *International and Comparative Law Quarterly* 423, 434.

24 Some commentators do not believe that this Act completely rules out the possibility of a future wrongful life action. See, for example, S. Deakin, A. Johnston and B. Markesinis, *Tort Law*, fifth edition, Oxford: Oxford University Press (2003), 309; J. Fortin, 'Is the 'Wrongful Life' Action Really Dead? 9 (1987) *Journal of Social Welfare Law*.

25 Congenital Disabilities (Civil Liability) Act 1976, s. 1 (2) (b).

26 *McKay v Essex AHA* [1982] QB 1166, Stephenson LJ; 1182, Ackner LJ, 1186–1187; Griffiths LJ, 1191.

27 S. 1ZA (1) (b), HFEA 1990.

comments will also be reviewed by the Licence Committee assessing the application.

Advice from peer reviewers and, where necessary, other organisations or groups will be sought to help the Authority determine the significant risk and seriousness of the disease/condition.[28]

This careful consideration of when life can be screened out sits awkwardly with the argument that life itself cannot be recognized as harm, on the basis that this would constitute an inroad into the sanctity of life.

Section 1 of the Congenital Disabilities (Civil Liability) Act 1976 was extended in 1990 to cover the infertility treatments regulated by the Human Fertilisation and Embryology Act. Section 1A states:

(1) In any case where –

   (a) a child carried by a woman as the result of the placing in her of an embryo ... is born disabled,

   (b) the disability results from an act or omission in the course of the selection, or the keeping or use outside the body, of the embryo carried by her or of the gametes used to bring about the creation of the embryo, and

   (c) a person is under this section answerable to the child in respect of the act or omission,

the child's disabilities are to be regarded as damage resulting from the wrongful act of that person and actionable accordingly at the suit of the child.

So the child's disability must *result* from an act or omission in the course of the selection, keeping, or use outside the body of the embryo. If the medical team somehow damage an embryo in the course of its selection, thereby causing the resulting child to suffer a physical manifestation which had not occurred on conception, the child will have a claim under s. 1A. However, the section provides for disability resulting from selection and it covers omissions in that process of selection. How a disability could *result* directly from an omission in the course of the selection is not clear. Where an embryonic defect was already present and cannot be cured, no disability could *result* from this. What could *result* from this is a disabled life. If it were possible to remedy defects in embryos prior to implantation, it might be possible to say

28   Available HTTP: www.hfea.gov.uk/5259.html (accessed 26 May 2010).

that where the medical team omits to check whether the particular embryo has been cured prior to implantation, the omission *results* in the disability, because the genetic defect would have been corrected before transfer. However, a failure to remedy a remediable defect in an embryo primarily amounts to an omission in the treatment of, rather than the selection of, the embryo. It is currently unlawful to implant in a woman an embryo whose nuclear or mitochondrial DNA has been altered or has had any cell added to it.[29] Thus, treating embryos is not possible. Furthermore, it is difficult to see how an originally healthy embryo could be harmed by an omission which could lead to the birth of a child who is harmed by that omission. Thus, the ambit of s. 1A is not clear.[30]

The USA: wrongful life, harm and the declining importance of the concept of sanctity of life

The Californian Court of Appeal has accepted that life itself can amount to harm.[31] In *Curlender v Bio-Science Laboratories*[32] the claimants were aware that their genetic heritage made them significantly more likely to conceive a child with Tay-Sachs disease,[33] so they approached their doctor to ascertain whether they were carriers of the gene before conceiving. The laboratory negligently provided erroneous information. As a result, the couple's doctor wrongly informed them that they were not carriers. In fact, both the Curlenders were carriers of recessive genes for Tay Sachs, which meant that there was a 25 per cent chance that their offspring would manifest the condition. Relying on the negligent advice, the Curlenders conceived Shauna, who had Tay-Sachs disease.

Shauna brought an action for wrongful life against the doctor and the laboratory. Prior to this, the US wrongful life cases had mostly concerned post-conception injuries similar to Mary McKay. These cases had met with disap-

29 S. 3ZA (4) (b) and (c).
30 Jane Miller, embryologist, University of Cardiff, is of the opinion that an originally healthy embryo cannot be harmed by an omission which could lead to the birth of a child who is in some way damaged by that omission. She suggested that an embryo might be harmed by an omission if there was a failure in its incubation, but that an embryo damaged in this way would not develop and, therefore, would not lead to the birth of a child who might then have an action under the Congenital Disabilities (Civil Liability) Act 1976. Personal communication.
31 As have courts in France and the Netherlands. *Assemblée plénière. Cour de Cassation,* pourvoi n.99–13, 701 (arrêt du 17 novembre 2000). Molenaar (March 26, 2003), Hetgerechtshof, Haag; However, the decision by the Cour de Cassation has been abrogated by a special statute after strong political lobbying by medical insurers. See, K. A. Warner, 'Wrongful Life Goes Down Under' 123 (2007), *Law Quarterly Review* 209, 210. See also Kirby J's strong dissent arguing that wrongful life actions should be recognized in the Australian case of *Harriton v Stephens* [2006] HCA 15.
32 *Curlender v Bio-Science Laboratories* 106 Cal. App. 3d 811, 165 Cal. Rptr. 477 (1980).
33 Tay Sachs disease is an inherited recessive condition that causes a progressive degeneration of the central nervous system, which is fatal (usually by age 5).

proval, on the basis that by being brought into existence the child has not suffered any damage cognizable at law.[34] Thus, in *Curlender* the crux of the problem was whether the child could be said to have suffered a harm.[35] The Californian Court of Appeal was of the opinion that there was no universal acceptance that the notion of religious beliefs, rather than law, should govern the situation.[36] Moreover, they suggested that considerations of public policy should include regard for social welfare as affected by careful genetic counselling and medical procedures.[37] The court felt that it would be unjust to 'retreat into meditation on the mysteries of life, and concluded that they need not be concerned with the fact that if the defendant had not been negligent Shauna might not have existed at all'.[38] Unconstrained by religious convictions concerning the sanctity of life, the court in *Curlender* concluded that the reality of the wrongful life concept was that the claimant both 'existed and suffered'.[39] Once harm had been recognized, given that the very reason for the defendant's undertaking was to prevent that harm, it followed that the defendant owed a duty to prevent that harm.[40] According to the court, the public policy considerations with respect to the individuals involved and to society as a whole, dictated recognition of a duty of care with respect to the harm of life in these circumstances.[41]

The Court of Appeal emphasized the fact that the predictive capability of modern genetics was such that genetic impairment was no longer a mystery.[42] Sonnenburg[43] argues that wrongful life actions arising from negligence in modern genetic techniques should be replaced with a tort of genetic malpractice.[44] On this analysis, the notion of wrongful life as 'harm'

---

34 See, for example, *Berman v Allan* 80 NJ 421, 404 A. 2d 8 (1979), Pashman J, 429–430; *Becker v Schwartz* 46 N.Y.2d 401(1978), Jasen J, 411; *Speck v Feingold* 408 A. 2d 496 (1979), Cercone J, 508; H. Teff, 'The Action for 'Wrongful Life' in England and the United States' 34(1985), *International and Comparative Law Quarterly* 423, 432. There is also a group of claims known as the 'status cases' where healthy, but illegitimate, children claim that they have been impaired by their illegitimate status. See, *Zepeda v Zepeda* 41 Ill. App. 2d 240 (1963); *Williams v New York* 18 N.Y. 2d 481 (1966); and *Stills v Gratton* 55 Cal. App. 3d 698 (1976).

35 *Curlender v Bio-Science Laboratories* 106 Cal. App. 3d 811, 165 Cal. Rptr. 477 (1980).

36 Ibid. Jefferson J, 486

37 Ibid. Jefferson J, 487–488.

38 Ibid. Jefferson J, 488.

39 Ibid.

40 The Curlender's stated that 'the very purpose for which the respondents' services were retained was to avoid conception' of an impaired child. Reply Brief for the Appellants, 10, in M. Sonnenburg, 'A Preference for Non-Existence: Wrongful Life and a Proposed Tort of Genetic Malpractice' 55 (1982), *Southern California Law Review* 477, 494.

41 *Curlender v Bio-Science Laboratories* 106 Cal. App. 3d 811, 165 Cal. Rptr. 477 (1980), Jefferson J, 488.

42 Ibid.

43 M. Sonnenburg, 'A Preference for Non-Existence: Wrongful Life and a Proposed Tort of Genetic Malpractice' 55 (1982), *Southern California Law Review* 477.

44 This is not the argument that is made here. Here the focus is on the ability of the tort of negligence, as inbued with an interest in personal autonomy to respond to the novel genetic claims.

is shaped by the ability of modern genetics to accurately predict and prevent genetic disease and disability.

Research demonstrates that there is a genetic component to most illnesses that affect human kind.[45] The increasing availability and acceptability of PGD and preconception genetic counselling may strengthen the view that a person has been harmed by being born.[46] Teff agrees, stating:

> Growing realisation of the predictive capabilities of genetic technology and its capacity to produce successful medical outcomes, as well as enhanced public awareness of the burdens of genetic disorders, have strengthened consumer demands for compensation when things go wrong.[47]

Many of the disorders for which PGD is authorized are associated with significant suffering.[48] Public sympathy is likely to run particularly high for people suffering from these kinds of conditions. Heightened awareness of the avoidability of genetic disease increases the likelihood that people will no longer be content to view abnormalities which might be prevented by modern screening techniques as fate.[49]

However, since *Curlender*, the Californian Supreme Court has refused to award general damages for wrongful life. In *Turpin v Sortini*,[50] the Turpins took their first child, Hope, to Sortini, a hearing specialist. He negligently informed them that Hope's hearing was within the normal limits when, in fact, she was totally deaf due to a hereditary ailment, which could affect any of the Turpin's offspring. The Turpins did not learn of Hope's problem until much later when she was examined by other hearing specialists. In the meantime, the Turpins conceived Joy, who was affected by the same hearing problem. Joy brought an action alleging that she had been deprived of the right to be born as a whole, functional human being without total deafness.[51]

---

45 The National Human Genome Research Institute states: 'We now believe that all diseases have a genetic component, whether inherited or resulting from the body's response to environmental stresses'. Available HTTP: www.genome.gov/10001191 (accessed 6 December 2010).

46 An increase in the availability of PGD will also, of course, lead to an increased potential for negligent mistakes and subsequent legal challenge.

47 H. Teff, 'The Action for 'Wrongful Life' in England and the United States' 34 (1985), *International and Comparative Law Quarterly* 423, 424.

48 Conditions such as Tay Sachs disease, Lesch Nyhan disease, cystic fibrosis and Huntington's chorea. The central list is fairly long; currently 139 conditions, many of which are associated with significant suffering. See the central list. Available HTTP: www.hfea.gov.uk/pgd-screening.html (accessed 11 May 2010).

49 H. Teff, 'The Action for "Wrongful Life" in England and the United States' 34 (1985), *International and Comparative Law Quarterly* 423, 424.

50 *Turpin v Sortini* (1982) 643 P 2d 954.

51 Framed this way the claim was problematic because, of course, Joy could never have existed without that genetic defect. The only way she might not have suffered that affliction was if she had never existed.

The court criticized *Curlender,* stating that it ignored the essential nature of the defendant's alleged wrong and obscured the critical difference between wrongful life actions and ordinary pre-natal injury cases. The Supreme Court held that Joy never had a chance to be born as a whole functional human being without total deafness, thus, her only complaint could have been that she should not have been born at all.[52] Relying on the pre-*Curlender* case law, the Supreme Court held that the claim for general damages should fail because it was impossible to determine in any rational or reasoned fashion whether the claimant had in fact suffered an injury in being born impaired, rather than not being born.[53] However, the court was prepared to award special damages for the extraordinary expenses for specialized teaching, training and hearing equipment. This conclusion suggests that the Supreme Court considered Joy's harm to be limited to the economic loss arising from her disability. This result is problematic because the Supreme Court held that Joy could not be said to have suffered any injury by being born impaired, rather than not being born. Nevertheless, they awarded the costs attributable to the disability with which she was born, despite the fact that the disability was in no way caused by the defendants.[54] This result precipitated a strong dissent from Mosk J, who felt that it was internally inconsistent to allow a child to recover special damages for a wrongful life action whilst denying general damages for the very same tort. He concluded that *Curlender* remained the prevailing law of California.[55]

In *Turpin,* the Supreme Court felt that legal decisions should not be based on public policy concerning the sanctity of life. Kaus J said:

> while our society and our legal system unquestionably place the highest value on all human life, we do not think it is accurate to suggest that this state's public policy establishes – as a matter of law – that under all circumstances 'impaired life' is preferable to 'nonlife'.[56]

The perceived minor nature of the genetic condition, which affected Joy Turpin, seemed to be a major factor in the court's decision not to recognize her 'impaired life' as worse than no life. The court opined that in this case where the claimant's only affliction was deafness, it was unlikely that a jury would conclude that life with such a condition was worse than not being born at all.[57] However, the court conceded that with respect to much more

---

52   *Turpin v Sortini* (1982) 643 P 2d 954, Kaus J, 961.
53   Ibid. Kaus J, 963.
54   On this distinction between types of harm, see I. Kennedy and A. Grubb, *Medical Law,* third edition, London: Butterworths (2000), 1548.
55   *Turpin v Sortini* (1982) 643 P 2d 954, Mosk J, 966.
56   Ibid. Kaus J, 961.
57   Ibid. Kaus J, 962.

serious debilitating and painful conditions, where the child has a very short life span and a very limited ability to perceive or enjoy the benefits of life, it could not assert with confidence that in every situation there would be a societal consensus that life is preferable to never having been born at all.[58]

This suggests that the seriousness of the person's disorder will have a role in determining whether she is deemed to have suffered an injury cognizable in law by being born with disabilities. In English law the notion of seriousness plays a significant role in the ability to terminate pregnancies and discard embryos for reasons of disability.[59] Opinions about the harmfulness of genetic conditions will vary; where the affliction is generally perceived to be mild, as the Californian Supreme Court suggested Joy's condition was, many might be unable to accept the person's argument that life with the particular condition is so harmful that it would have been preferable that she was not born. However, where the condition is degenerative and associated with significant suffering, there might be more support for the person's wish not to have been born. Although, at first glance, an approach to recognizing the harm in wrongful life claims based on the severity of the disorder might appeal to public sympathies, relying on this as the basis for a legal approach will create problems when it comes to drawing the boundaries of liability.

By virtue of the power conferred on it by s. 1 ZA (1) (b), the Human Fertilisation and Embryology Authority is progressively relaxing the criteria upon which embryos might be deselected, so that embryos with lower penetrance genetic disorders, which merely increase the risk that that person will suffer a particular condition later in life, might be screened out. Where negligence in performing PGD leads to the birth of a child with the BRCA1 gene, who can legitimately be screened out under Schedule 2, s. 1 ZA of the HFE Act, that child may never manifest the genetic condition associated with that genetic defect; namely breast or ovarian cancer.[60] Even if she does manifest that condition, she is not likely to do so until later in life.[61] However, if the legitimacy of screening out the 'serious' disorders recognized

---

58 Ibid. Kaus J, 963. For more on how courts might make distinctions based on severity with respect to wrongful life cases, see the discussion below.

59 See the Abortion Act 1967, s. 1 (1) (d), whereby abortion is permitted if 'there is a substantial risk that if the child were born it would suffer from such physical or mental abnormalities as to be seriously handicapped'. See also the Human Fertilisation and Embryology Act, Schedule 2, s. 1ZA (2) (b), which permits the testing of embryos (for the purpose of de-selection) where there is a 'significant risk that a person with the abnormality will have or develop a serious physical or mental disability, a serious illness or any other serious medical condition'.

60 Perhaps because she avoids environmental factors that may play a part in the manifestation of the condition.

61 This ignores any argument that the child might make about living a life blighted by the knowledge that she is likely, one day, to manifest a particular condition, because similar concerns are considered in Chapter 7 when analysing the right not to know.

by the HFEA[62] is the basis for the recognition that life with a genetic disorder can amount to harm, it would seem to follow that any person with respect to whom a legitimate decision[63] to screen out was made, could demonstrate harm. Although, theoretically speaking, it might be possible to sustain a wholesale approach to the harm of being born whenever the embryo from which the individual with the genetic condition developed could have been screened out, legally speaking, the courts might want to exercise a greater degree of control over the extent of liability which exceeds the legitimacy of the reasons for screening out. Although negligence leading to the type of claim considered here is unlikely to be commonplace, thereby generating a significant number of claims, the courts might want to protect public authority defendants from costly claims.[64] Thus, the courts might prefer an approach whereby they determine whether a life with a genetic condition (which could have been legitimately prevented via PGD) could be harm, which is independent of the HFEA's decision on the legitimacy of screening out per se.[65]

An interest in autonomy as the basis for the legal recognition of harm in negligently caused life

The principle of the sanctity of life is at odds with any notion of variance based on quality, quantity or willingness. In the context of the crimes of murder or manslaughter, the notion of sanctity of life is an overriding general principle which leads to the view that taking life is intrinsically wrong.[66] The specifics of the taken life are irrelevant.[67] However, in English medical law, the nature of the specific life is crucial to the question of whether the life is sacred. In medical law the principle of sanctity of life is not an inalienable principle; it is questioned and analysed in the context of particular lives.

---

62 See s. 1ZA (2) (b) and the central list. Available HTTP: www.hfea.gov.uk/pgd-screening.html (accessed 11 May 2010).

63 On the basis of the HFEA's regulatory position.

64 The Department of Health has confirmed that PGD is available on the NHS but, currently, there are significant discrepancies between primary care trust's funding policies. See D. Rose, 'Embryo Screening Funding is a "Postcode Lottery" ', *The Times* (Jan. 15 2009). For an NHS decision to fund treatment, see *Bionews*, 'NHS to Fund Saviour Sibling Treatment' (22 March 2004). Available HTTP: www.bionews.org.uk/page_11894.asp (accessed 16 June 2010).

65 The HFE Act and the central list of disorders for which a licence to screen out embryos can be provided would be the starting point.

66 H. T. Engelhardt, *The Foundations of Bioethics,* Oxford: Oxford University Press (1986), 9.

67 There is no defence for mercy killing in English law. The intentional killing of a terminally ill patient would be murder. However, this does not mean that juries are not lenient in such circumstances.

Determining the value which ought to be attached to human life with a view to identifying the circumstances, if any, when human life might be violated, is one which regularly arises in the context of making decisions about medical treatment.[68] Many philosophers do not regard all life as of ultimate value. Kant saw rational beings as of inherent value, not all beings.[69] From this perspective, rationality is what imbues life with value. In a similar vein, Ronald Dworkin adopts a position which emphasizes the notion of identity as making life valuable. He argues that 'life's inherent value ... depends on the intrinsic importance of human creative investment in it. On this view it is not biological life that is sacred but human life created by personal choice, training, commitment and decision.'[70]

On this view, life does not have ultimate or equal application; it is not valuable in, and of, its biological self. Its value lies in its qualities. There must be a life for the qualities to attach to, but it is these qualities that imbue the life with ultimate value.

English medical law: autonomy and the implications for the principle of sanctity of life

In general, modern medicine seeks to promote life. Historically, adhering to a principle where life should be regarded as sacred and preserved at all costs met with little opposition because there was little that could be done to prevent death in any event. Modern medical technology can sometimes maintain life for periods beyond its natural end. This ability has brought the question of when life should be regarded as sacred, and therefore prolonged, into sharp focus.

Airedale NHS Trust v Bland represents one of the most important English cases concerning the sanctity of life. In the House of Lords, Lord Goff affirmed the importance of the principle: 'The fundamental principle is the principle of the sanctity of human life – a principle long recognized not only in our own society but in most, if not all, civilized societies throughout the modern world.'[71] However, he continued: 'But this principle, fundamental as it is, is not absolute.'[72]

Thus, in medical law, the principle of sanctity of life is capable of being overridden. The problem is in defining the circumstances when an exception

---

68 See, for example, *Airedale NHS Trust v Bland* [1993] AC 789 and a number of cases involving infants: *Re B* [1990] All ER 927; *Re C (A Minor)* [1990] Fam 26; *Re J (A Minor)* [1991] Fam 33; *Re C (A Minor)* [1998] 1 FLR 384; *A NHS Trust v D* [2000] 2 FLR 677.

69 For comment, see T. Hayward, 'Kant and the Moral Considerability of Non-Rational Beings' 36 (1994), *Royal Institute of Philosophy Supplement* 129.

70 R. Dworkin, *Life's Dominion,* London: Harper Collins (1993), 157.

71 *Airdale NHS Trust v Bland* [1993] AC 798, Lord Goff, 864.

72 Ibid.

to such a fundamentally important principle should be made. The forbearance towards the principle of the sanctity of life in medical law has diminished in recent years as the importance of the principle of autonomy has risen. In the face of the withdrawal of life-sustaining treatment, the individual's will is particularly important in determining the relative importance of the two principles.

Where the individual can make her own decisions, the law sometimes allows her to make her own assessment of the quality of her life and we respect that assessment because we respect her autonomy.[73] Where a life might be truncated by the omission to provide relevant medical care, respect for autonomy overrides the position that life is sacred. Lord Donaldson's dicta in *Re T* sums up the position well: 'An adult patient who … suffers from no mental incapacity has an absolute right to choose whether to consent or to refuse medical treatment, even if it will damage his health or lead to his premature death'.[74]

The emphasis on the principle of autonomy at the expense of other principles relevant in medicine is well documented.[75] Its supremacy over the principle of sanctity of life in the context of withdrawing and withholding treatment in English law can be seen in *Re B (Consent to Treatment: Capacity)*.[76] Ms B suffered an event that rendered her tetraplegic. As a result, she was kept alive by artificial ventilation. Although she had capacity, her repeated requests that the artificial ventilation be withdrawn were not respected. Ms B challenged the lawfulness of the treatment against her will. Dame Butler-Sloss thought that the principle of self-determination must prevail over the principle of sanctity of life.[77] Currently, the respect afforded to the principle of autonomy at the expense of the principle of sanctity of life, is largely confined to the withholding or withdrawal of treatment. However, recently there appears to be a growing willingness, both legally and generally, to recognize that an inroad may be made into the sanctity of life in circumstances where death requires a treatment intervention rather than a removal of treatment.

In English law, assisted suicide is criminalized by s. 2 (1) of the Suicide Act 1961. Two fairly recent cases have challenged the provisions on assisted

---

73 *Re B (Consent to Treatment: Capacity)* [2002] 1 FLR 1090. The lack of similar cases following this one perhaps suggests that health professionals largely respect autonomous patient's wishes not to continue with life-sustaining medical treatment without recourse to the courts.

74 *Re T (Adult: Refusal of Treatment)* [1993] Fam 95, Lord Donaldson, 102.

75 See, for example, C. Foster, *Choosing Life, Choosing Death: The Tyranny of Autonomy in Medical Ethics and Law,* Oxford: Hart (2009); O. O'Neil, *Autonomy and Trust in Bioethics,* Cambridge: Cambridge University Press (2002); G. Laurie, *Genetic Privacy; A Challenge to Medico-Legal Norms,* Cambridge: Cambridge University Press (2002).

76 *Re B (Consent to Treatment: Capacity)* [2002] 1 FLR 1090.

77 Ibid. Dame Butler-Sloss, 1096–1098.

suicide on human rights grounds. In 2001 Diane Pretty unsuccessfully challenged the Director of Public Prosecution's (DPP) refusal to give assurance that her husband would not be prosecuted under s. 2 of the Suicide Act if he helped her to commit suicide. Drawing on this claim, Debbie Purdy brought a similar action. Purdy has multiple sclerosis, a debilitating and progressive illness, and believes that at some point she will find her life unbearable and will want to travel to Switzerland where assisted suicide is legal. She wants to be able to delay travel until she is no longer able to make travel arrangements herself. Consequently, she wants to rely on her husband to assist her travel. Purdy is anxious that in doing so her husband risks prosecution under s. 2 (1) of the Suicide Act. Therefore, she brought an action for judicial review against the DPP on the basis that the failure to give specific advice as to when a prosecution might be brought for assisting suicide was a breach of her rights under Article 8 of the European Convention on Human Rights.

According to the House of Lords, Article 8 was engaged in this case.[78] The House allowed Purdy's appeal and required the DPP to provide an offence-specific policy identifying the facts and circumstances which he will take into account in deciding whether or not to consent to a prosecution under s. 2 (1) of the 1961 Act.[79] Although the House did not expressly comment on what the content of that policy should be, this decision suggests that the House believed that assisting an autonomous person's suicide ought not automatically to lead to prosecution.

Baroness Hale said 'the object must be to protect the right to exercise a genuinely autonomous choice. The factors which tell for and against such a genuine exercise of autonomy free from pressure will be the most important.'[80] In response to this, the DPP consulted with over 5,000 individuals and groups, and published a policy providing a framework for prosecutors to decide which cases should proceed to court and which should not.[81]

Indirectly, the House of Lords' judgment in favour of Mrs Purdy, the public consultation and the subsequent policy on assisted suicide, demonstrate a growing acceptance that the principle of sanctity of life should not prevent an autonomous desire to be assisted in committing suicide. The first, and perhaps the foremost, of the factors against prosecution in the DPP's guidance is that the victim had reached a voluntary, clear, settled and informed decision to commit suicide.[82] Similarly, two of the primary factors in favour of prosecution are that the victim had not reached a voluntary, clear, settled and informed decision to commit suicide and that the victim

---

78 *R (on the application of Purdy) v DPP* [2009] UKHL 45, Lord Hope, 39.

79 Ibid. Lord Hope, 56.

80 Ibid. Baroness Hale, 65.

81 Crown Prosecution Service, *DPP Publishes Assisted Suicide Policy* (25/02/2010). Available HTTP: www.cps.gov.uk/news/press_releases/109_10/ (accessed 21 June 2010).

82 Ibid.

had not clearly and unequivocally communicated his or her decision to commit suicide to the suspect.[83] If the only thing that is standing in the way of an individual's clear and settled wish to commit suicide is that she needs assistance, and she is reluctant to ask for that assistance where it will put the assistor at risk of prosecution, then these guidelines will in some circumstances serve as protection for autonomy over the principle of sanctity of life. Interestingly, this indirect protection for the autonomous wish to die is not limited to those who are terminally ill. In the context of voluntary lethal treatment, which the Assisted Dying for the Terminally Ill Bill sought to legalize in 2002, the suggestion was that any legal right would be confined to protecting the autonomous wish to die of the terminally ill only.[84] It might be argued that not limiting the assisted suicide non-prosecution policy to the terminally ill stands to make a significant inroad into the principle of sanctity of life because it will make it easier for a person to fulfil her settled wish to die where she is not dying and could have lived a long life.[85]

Application to wrongful life claims

These developments call into question the Court of Appeal's refusal in *McKay* to recognize that an individual could bring a claim based on the premise that she should not have been born because this would make an inroad into the principle that life is sacred.[86] Mary McKay was only six years old when the claim was brought on her behalf and, thus, was not capable of expressing a view on the worthwhileness of her life.[87] The Human Fertilisation and Embryology Authority continues to issue licences for screening out particular genetic disorders. Currently, the Authority is adopting an expansive approach to the issuing of licenses for excluding embryos with certain genetic conditions.[88] The list contains conditions which one might expect where death comes in early childhood and the condition is associated with significant pain and suffering.[89] However, the list also contains a number of conditions which might not in fact manifest because they do not have 100 per cent penetrance, and even if they do manifest, it will be in adulthood. For example, it is legitimate to screen out embryos which have the BRCA 1 genetic mutation and are therefore more susceptible to adult onset breast

---

83   Ibid.

84   See the Assisted Dying for the Terminally Ill Bill, s. 2 (2) (c) and (d), tabled by Lord Joffe in 2002.

85   If quantity is seen as central to the concept of sanctity of life.

86   *McKay v Essex AHA* [1982] 1166 QB Stephenson LJ, 1180; Ackner LJ, 1188.

87   She may, of course, never have been capable of this given the extent of her mental disabilities.

88   The HFEA list currently consists of 139 conditions. Available HTTP: www.hfea.gov.uk/pgd-screening.html (accessed 11 May 2010).

89   Some examples of such conditions on the list might be Lesch Nyhan Syndrome and Tay Sachs disease.

cancer.[90] The increasing ability to screen out embryos which are susceptible to late onset disorders which do not interfere with intellectual capability, creates greater scope for the created person to reach a point in life where she is able to make a personal assessment of whether it would have been better if her parent's reproductive project had succeeded and she had never been born. Furthermore, even where the particular genetic condition manifests in childhood, many people affected by genetic conditions are living into adulthood as a result of improved medical management and, therefore, may become capable of forming a view on whether they wish that they had never been born.[91] Although some of the conditions which can be screened out will interfere with intellectual capacity, such that the sufferer will not attain the ability to comprehend that negligence led to her birth and make a subsequent assessment about whether or not she is pleased that she was born, many of the conditions, which can legitimately be screened out, do not interfere with intellectual capacity.[92]

Assume that potential parents seek to screen out embryos which have a childhood onset condition which does not necessarily lead to death in early childhood, or an adult onset condition. Assume further that the parent's project is frustrated due to lack of due care and attention so that a child with the deleterious gene is born. Assume further that upon maturity, the person who would have been screened out wishes that her parent's project had been successful and that she had never been born.[93] Where the claim that no life would have been better than the life led is articulated by the person experiencing that life, should the claim be rejected on the basis that it would be an inroad into the sanctity of life?

The difficulty with the argument that the individual's autonomy is interfered with, where she was born against the legitimate decision of her parents and she subsequently endorses the decision they made on her behalf, is that the individual who eventually comes into existence did not have a direct choice in the matter. Currently, the decision that it would be better for some people not to be born and, thus, that some types of embryos should be screened out, is based on an objective welfare assessment, as opposed to a subjective assessment of what the potential individual might want.[94]

---

90 Similarly, parents can screen out FAP, a hereditary bowel cancer, and early onset Alzheimer's Disease.

91 Cystic fibrosis is the most common Mendelian disorder of children and young adults of caucasian descent. It is, therefore, one of the most common disorders screened out via PGD in the UK. Life expectancy was under 5 in the 1960s, but is around 40 in the present day.

92 For example, cystic fibrosis, polycystic kidney disease and Becker's muscular dystrophy.

93 Because life with her disabilities, or in the shadow of the impending prospect of fatal genetic illness, is unbearable.

94 It is argued below that the latter could arise from an autonomy-based position.

Although welfare is a central element of decisions to provide assisted conception,[95] Emily Jackson convincingly argues that its consideration at the pre-
conception stage is incoherent and irrelevant.[96] The welfare of future children
is central to the decision to provide any assisted conception services as evidenced by s. 13 (5) of the Human Fertilisation and Embryology Act 1990,
which states: 'A woman shall not be provided with treatment services unless
account has been taken of the welfare of any child who may be born as a result
of the treatment (including the need of that child for supportive parenting).'
Determining the basis of the decision to not select a particular embryo, and,
by corollary, a particular person, is important for the purposes of this chapter
because the interest upon which the original decision is based determines the
interest which the carelessness interferes with. As argued here, it seems that
the basis for the decision not to select a particular embryo under the Human
and Fertilisation and Embryology Act, is that it would be contrary to the
welfare of that future individual that she be born. This was essentially the
grounds upon which Mary McKay's claim was made. That is, that living with
the level of suffering which she did was contrary to, and amounted to, an interference with her welfare.[97] However, her argument that her life was contrary to
her welfare was resoundingly rejected by the Court of Appeal.

Pre-conception evaluation of both welfare and autonomy are problematic because we cannot know what that individual will perceive as suffering or whether she would have chosen to be born. However, it might be
argued that empathetic judgments about a future person's welfare are more
easily made than empathetic judgments about what a future person would
have wanted. Although some philosophers adopt subjective theories of
welfare,[98] it might be argued that it can be objectively determined what
welfare consists of more easily than it can be determined what a person's
desires consist of.[99] Drawing on the Aristotelian theory of objective
welfare, Nussbaum produces a list of ten objective capabilities that are
crucial to welfare.[100] In the medical context, Anand argues that welfare is

95   See s. 13 (5) Human Fertilisation and Embryology Act and discussion below.

96   E. Jackson, 'Conception and the Irrelevance of the Welfare Principle' 65 (2002), *Modern Law Review*
176.

97   In *McKay v Essex AHA* [1982] QB 1166, both Stephenson LJ, 1184, and Ackner LJ, 1189, acknowledged that Mary was enduring a level of suffering and if the defendants had not been careless she
would not be suffering now because she would not be alive.

98   See, for example, L. W. Sumner, *Welfare, Happiness and Ethics,* Oxford: Clarendon Press (1996).

99   In the context of medical ethics the concept of welfare has received significant consideration. See, for
example, J. Harris, *The Value of Life: An Introduction to Medical Ethics,* London and New York: Routledge
(1985); T. L. Beauchamp and J. F. Childress, *Principles of Biomedical Ethics,* fifth edition, Oxford: Oxford
University Press (2001). However, as Molynoux notes: 'many medical ethicists, while writing about
welfare, are vague about what they mean by this word'. D. Molynoux, '"And how is life going for you?"
An Account of Subjective Welfare in Medicine' 33 (2007), *Journal of Medical Ethics* 568.

100   M. Nussbaum, 'Non-relative Virtues, an Aristotelian Approach' in A. Sen, M. Nussbaum (eds), *The
Quality of Life,* Oxford: Clarendon Press (1993), 242.

best seen in terms of objective functionings and capabilities.[101] In the context of its legal conception, autonomy tends to be a more subjective notion which amounts to simply doing as one pleases.[102] However, it might be argued that a person always exercises her autonomy in line with her own welfare. If autonomy and welfare are intertwined,[103] that is, if the question of your welfare consists of freely chosen goals, desires and actions, it might be argued that autonomy becomes less subjective and welfare becomes more subjective.[104] However, when one moves towards a more substantive or value-based account of autonomy,[105] it becomes easier to speak objectively about what autonomy consists of. The concept develops some characteristics which can be externally evaluated. In particular, accounts of autonomy which are based on substantive accounts of rationality may enable analytical purchase, as might those procedural accounts which require demonstration of an ability to critically reflect on first order desires at a higher order.[106]

Thus, let us consider whether it is possible to analyse the decision made by the parents in terms of the autonomy of the subject of their decision, with a view to considering whether the individual might argue that the careless failure to respect that decision amounts to an interference with her autonomy. If the careless failure to respect the parents' decision not to select a particular embryo can conceivably be interpreted as an interference with the created individual's autonomy, it must be because the parental decision amounts to a substituted judgment, rather than, as currently appears to be the case, a decision which is concerned with the welfare of the future child and is, therefore, based on best interests.

101 P. Anand, 'Capabilities and Health' 31 (2005), *Journal of Medical Ethics* 299. Harris and Daniels also describe welfare in terms of needs that are 'objectively ascribable; we can ascribe them to a person even if he does not realize he has them and even if he denies he has them because his preferences run contrary to the ascribed needs'. J. Harris and N. Daniels, 'Health Care Needs and Distributive Justice' in: J. Harris (ed.), *Bioethics: Oxford Readings in Bioethics,* Oxford: Oxford University Press (2001), 319–47. This discussion does not attempt to provide an account of what objective welfare consists of. It argues that welfare can be construed as objective. But the focus of this discussion is the ability of autonomy, as opposed to welfare, to provide a basis for decisions to screen out embryos.

102 This view represents the liberal interpretation of autonomy as consisting of procedurally independent subjective preferences and desires which, as argued throughout this book, represent the interpretation which the English courts currently purport to adopt with regard to trespass actions relating to consent to medical treatment.

103 As it is argued they are in Chapter 3. See, in particular, Raz's views on the connection between autonomy and well being J. Raz, *The Morality of Freedom,* New York: Oxford University Press (1988), Chapter 14.

104 This is a difficult philosophical question, which I do not intend to conduct a more detailed inquiry into here.

105 See Chapter 3 for a discussion of these types of autonomy.

106 See the discussion of value and procedural rationality in Chapter 3.

Making decisions requires a certain level of capacity.[107] Not least, the decision-maker must be able to understand the information relevant to the decision. Where she is incapable of this, another may have to make a vicarious decision on her behalf. Two models commonly guide the process of vicarious decision-making; substituted judgment, which is prevalent in the United States, and best interests, which is the standard usually used in the UK.[108] As the welfare approach was unsuccessful in *McKay*, let us analyse how the hypothetical claim might look if the basis for the parents' decision not to select this person were deemed to be that of substituted judgment.

The test of substituted judgment provides an autonomy-based approach to the treatment of incompetent individuals.[109] The argument is that we can respect autonomy by following or reconstructing, as best we can, the autonomous decision a person would have made if she were able to.[110] The doctrine emerged in the US with regard to the question of gifts to be made from an incompetent's estate to one to whom the incompetent owed no duty of support.[111] In the US the approach has gained significant popularity among the judiciary with regard to medical treatment of incompetent patients,[112] leading one commentator to state: 'Substituted judgment remains the guiding framework for surrogate decision making in both bioethics and law.'[113]

---

107 For the legal requirements necessary to be able to make one's own decisions, see s. 3 of the Mental Capacity Act 2005.

108 As we know, the best interests standard is also the basis for decision-making under the HFE Act. In assisted conception services generally, the welfare assessment is made by the medical team who are considering whether to treat the potential parents. However, with respect to embryo testing, it might be argued that the decision regarding the welfare of a child with the condition in question rests with the parents within the context of the welfare structure set up by the HFE Act and the Authority which manages the central list. The treating clinic will still make a welfare assessment in the first place regarding whether the parents can access treatment per se, but the actual decision of whether it would be contrary to a potential child's welfare to bring her into existence with a particular genetic disorder will rest with the parents.

109 P. Lewis, 'Medical Treatment of Dementia Patients at the End of Life: Can the Law Accommodate the Personal Identity and Welfare problems?' 13 (2006), *European Journal of Health Law* 219, 219. See also D. Egonsson, 'Some Comments on the Substituted Judgement Standard' 13 (2010), *Medicine, Health Care and Philosophy* 33.

110 A. Jaworska, 'Advance Directive and Substitute Decision Making' (2009), *Stanford Encyclopedia of Philosophy*. Available HTTP: http://plato.stanford.edu/entries/advance-directives/ (accessed 17 May 2010).

111 See *Re Guardianship of Brice* 233 Iowa 183, 8 N. W. 2d 576 and *Re Buckley's Estate* 330 Mich. 102 47 N. W. 2d 33.

112 For some notable examples, see *Strunk v Strunk* (1969) 445 S. W. 2d 145; *Re Boyd* 403 A 2d 744 (DC CA 1979) Ferren LJ, 750; *In the Matter of Karen Quinlan* 70 N. J. 10 355 A. 2d 647 (1976); *Superintendent of Belchertown State School v Saikewicz* 373 Mass 728, 320 N. E. 2d 417 (1977). See also T. G. Gutheil and P. S. Appelbaum, 'Substituted Judgment: Best Interests in Disguise' 13 (1983), *The Hastings Center Report* 8, 8, for an account of the popularity of the substituted judgment doctrine in US medical law.

113 A. M. Torke, G. C. Alexander and J. Lantos, 'Substituted Judgment: the Limitations of Autonomy in Surrogate Decision Making' 23 (2008), *Journal of General Internal Medicine* 1514.

The aim of the approach is to decide vicariously what the subject of the decision would have decided if she could decide for herself.[114] In *Re Boyd*, The District of Columbia Court of Appeals said that in making a substituted judgment the duty of the court 'as a surrogate for the incompetent, is to determine as best it can what choice that individual, if competent, would make with respect to medical procedures'.[115]

However, some commentators doubt the ability of decisions made on behalf of individuals who cannot make their own decisions to protect the autonomy of the subject of the decision because the person purportedly making the vicarious decision for the incompetent subject cannot possibly avoid being motivated by best interest considerations, or by her own feelings.[116] Other commentators take a more nuanced view whereby it is possible to protect the autonomy of persons who cannot make their own decisions in certain circumstances. Lewis argues that if the incompetent individual was previously competent, her earlier autonomous decisions can be projected into the future once she becomes incompetent.[117] Explicit decisions regarding particular circumstances which are made by the person in advance are usually justified by recourse to autonomy on the basis that they protect the decision that the patient would have made.[118] However, even if the individual has not made an advance directive, it is possible that 'her autonomy can be respected by making the decision that she would have

---

114   T. G. Gutheil and P. S.Appelbaum, 'Substituted Judgment: Best Interests in Disguise' 13 (1983), *The Hastings Center Report* 8.

115   *Re Boyd* 403 A 2d 744 (DC CA 1979).

116   T. G. Gutheil and P. S. Appelbaum, 'Substituted Judgment: Best Interests in Disguise' 13 (1983), *The Hastings Center Report* 8, 9. See also A. M. Torke, G. C. Alexander and J. Lantos, 'Substituted Judgment: the Limitations of Autonomy in Surrogate Decision Making' 23 (2008), *Journal of General Internal Medicine* 1514; S. Bailey, 'Decision Making in Health Care: Limitations of the Substituted Judgment Principle' 9 (2002), *Nursing Ethics* 483.

117   P. Lewis, 'Medical Treatment of Dementia Patients at the End of Life: Can the Law Accommodate the Personal Identity and Welfare Problems?' 13 (2006), *European Journal of Health Law* 219, 219–220.

118   However, even where there is a specific and explicit advance decision, the justification on the basis of autonomy is not without contention. The validity of relying on advance decisions as expressions of self-determination has been criticized from the perspective of the personal identity objection. According to this objection, a person who makes an autonomous decision when competent, does not share her identity with the incompetent person to whom the choice then purports to apply. See, for example, A. Buchanan, 'Advance Directives and the Personal Identity Problem' 17 (1988), *Philosophy and Public Affairs* 277; D. Parfit, *Reasons and Persons*, Oxford: Clarendon Press (1984), 204–209; R. Dresser, 'Life, Death and Incompetent Patients: Conceptual Infirmities and Hidden Values in the Law' 28 (1986), *Arizona Law Review* 373, 379–381; P. Lewis, 'Medical Treatment of Dementia Patients at the End of Life: Can the Law Accommodate the Personal Identity and Welfare Problems?' 13 (2006), *European Journal of Health Law* 219, 221. However, some notable commentators reject the argument that the personal identity problem prevents advance directives being interpreted as expressions of self-determination. See, in particular, R. Dworkin, *Life's Dominion: An Argument about Abortion and Euthanasia*, London: Harper Collins (1993), 180–8, 210–16.

made, based on evidence of her previously competent wishes, preferences and values'.[119] This approach requires the vicarious decision-maker to be able to subjectively analyse what the individual might have wanted. Thus, it requires evidence indicating what the person would have wanted. Where there is no evidence of what the individual would have decided because she has always been incompetent, or she is a young child, some commentators argue that it is not possible to protect her autonomy via the substituted judgment approach.[120] However, other commentators adopt a thicker version of substituted judgment, which has an objective basis. From this perspective, it is possible to make substituted judgments for those for whom specific evidence of previous desires is not available. It is difficult to argue that autonomy on a basic, liberal conception is protected by the doctrine of substituted judgment.[121] However, where autonomy is imbued with a notion of substantive rationality, it is possible to see how objective substituted judgments might be made in the name of autonomy.

In the District of Columbia Court of Appeals in *Re AC*,[122] Terry JA said of substituted judgment:

> After considering the patient's statements, if any, the previous medical decisions of the patient, and the values held by the patient, the court may still be unsure about what course the patient would choose. In such circumstances the court may supplement its knowledge about the patient by determining what most persons would be likely to do in a similar situation.[123]

This approach suggests that what reasonable people would want in certain circumstances reflects a wider view of what people, including the particular subject of the decision in question, would want on the basis of rationality.[124] The notion of objective substituted judgment is considered by Robertson who notes that:

---

119   P. Lewis, 'Medical Treatment of Dementia Patients at the End of Life: Can the Law Accommodate the Personal Identity and Welfare Problems?' 13 (2006), *European Journal of Health Law* 219, 220.

120   See, in particular, P. Lewis, 'Medical Treatment of Dementia Patients at the End of Life: Can the Law Accommodate the Personal Identity and Welfare Problems?' 13 (2006), *European Journal of Health Law* 219, 219–220.

121   They believe that in these circumstances an objective account of the doctrine of substituted judgment collapses into the notion of best interests. See S. Pattinson, *Medical Law and Ethics*, second edition, London: Sweet and Maxwell (2009), 163.

122   *Re AC* 573 A. 2d 1235 (DC CA) (1990).

123   Ibid. Terry JA, 1251.

124   The perspective of the reasonable person as a basis for determining what rationality consists of, in a legal sense, is discussed at various points throughout this book. See, in particular Chapter 3.

The courts have had to choose between a subjective and an objective standard, the first requiring an actual indication of donative intent to the recipient, the second focusing on what the incompetent would be wise, prudent or reasonable to do.[125]

Robertson asks if 'no evidence of the (incompetent individual's) wishes exists, then should it not be assumed that he would act as a similarly situated, reasonable person?'[126] Rawls appears to acknowledge that desires can be imputed to individuals and simulated decisions made on their behalf on the basis of rationality. When addressing the issue of making decisions on behalf of others in a Theory of Justice he states: 'We must choose for others as we believe they would choose for themselves if they were at the age of reason and were deciding rationally.'[127] According to Robertson, the substituted judgment doctrine as objectively understood 'seeks to treat incompetents as competents are treated – as creatures of choice, with the autonomy and dignity of choice'.[128]

The argument that the interest interfered with in the hypothetical claim discussed in this chapter is one of autonomy, rests on the original parental decision being interpreted as one of substituted judgment as a protection of autonomy, rather than a decision which is based on welfare.[129] If, as suggested above, the substituted judgment doctrine does have an objective element, whereby the decision purports to simulate what the incompetent would want by reference to what she would be wise, prudent or reasonable to do,[130] it might be argued that the framework established by the Human Fertilisation and Embryology Act provides a clear objective basis of the circumstances into which people might not want to be born.[131] Acting within these guidelines, the parents arguably seek to simulate the evaluation of their potential child's reaction to a failure not to select her on the

---

125 J. A. Robertson, 'Organ Donations by Incompetents and the Substituted Judgment Doctrine' 76 (1976), *Columbia Law Review* 48, 59.

126 Ibid. 61.

127 J. Rawls, *A Theory of Justice,* Cambridge, MA: Harvard University Press (1971), 209. He does not base this definition of what I interpret to be substituted judgment on the principle of autonomy.

128 J. A. Robertson, 'Organ Donations by Incompetents and the Substituted Judgment Doctrine' 76 (1976), *Columbia Law Review* 48, 76.

129 As it is argued above, this represents the current basis for parental decision-making with respect to embryo de-selection within the context of the framework established by the Human Fertilisation and Embryology Act 1990.

130 J. A. Robertson, 'Organ Donations by Incompetents and the Substituted Judgment Doctrine' 76 (1976), *Columbia Law Review* 48, 59.

131 The Human Fertilisation and Embryology Act set up the Human Fertilisation and Embryology Authority which is charged with deciding when licences can be granted to screen out embryos. Part of the process of deciding whether a licence will be granted involves public consultation. See discussion earlier in this chapter.

basis of the particular disability and, upon this simulated evaluation of life with this condition, they decide that reasonable people and, therefore, their future child would rather not be born. There is a link here between autonomy and welfare.[132] Most of the time, most of us exercise our autonomy with the aim of maximizing our welfare.[133] As we know, Raz argues that well being enshrines the ideal of personal autonomy.[134] Although it can be argued that an objective account of the doctrine of substituted judgment collapses into the notion of best interests,[135] if the objective argument can be sustained, it might be argued that the position of autonomy as the basis for the decision in the hypothetical scenario considered becomes stronger in the face of the subject's autonomous evaluation and endorsement of the decision.

As the discussion above demonstrates, subjective evidence of previous desires enables protection of autonomy via the application of the doctrine of substituted judgement. This subjective element of the decision presupposes a subject,[136] which is a problem in the context of the hypothetical claim, because at the time of the parental decision there is no subject whose desires can be adduced. Considerations of substituted judgment which insist on subjectivity focus on the subject of the decision prior to the circumstances in which the decision is made. Thus, the decision can be deemed to be a protection of an incapacitated person's autonomy even if there is no way of discovering whether this is what the individual actually wanted. Even though past indication of subjective desires is not available in the hypothetical scenario, it might be argued that where the subject reaches a position whereby she is able to reflect on and either endorse or refute the decision made on her behalf, there is a greater ability for assessing whether the substituted decision does in fact reflect the subject's autonomy, than there is when only prior indication is available, and there is no chance that the desires implemented in the substituted decision could be evaluated and endorsed, or not endorsed by the subject. Subsequent evaluation arguably forms a clearer basis of the person's actual desires than previous indications because it injects a level of specificity not present in decisions made on the basis of prior indications. Where the subject is evaluating the decision, she is directing her mind to the precise issue in question and making a specific evaluation of whether this is what

---

132   This leads back to the discussion in Chapter 3 regarding the wider conception of autonomy as significantly more than a decisional concept.

133   At times this might, of course, be done via the maximization of the welfare of others or through the experience of altruism.

134   J. Raz, *The Morality of Freedom,* New York: Oxford University Press (1988), Chapter 14, 369. See the discussion on the link between autonomy and well being in Chapter 3.

135   S. Pattinson, *Medical Law and Ethics,* second edition, London: Sweet and Maxwell (2009), 163.

136   I would like to thank Dimitrios Kyritsis for elucidating this point to me and for the discussions which he engaged me in regarding this chapter.

she wants.[137] Even those who adopt the thin approach to subjective judgment as a reflection of autonomy hold that indications of previous wishes, preferences and values, as opposed to only specific advance directives, can enable the making of decisions which protect the subject's autonomy.[138]

The crux of the hypothetical case then, in terms of the autonomy analysis, is that the subject of the original decision[139] can evaluate the simulated decision and endorse it as a decision which purported to, and in fact protected, her desires. That is, the parents simulate the potential child's evaluation of the decision[140] and make a substituted judgment on her behalf, and when she becomes capable of autonomy, she evaluates that decision and endorses it as a reflection of her desires.[141] Both the autonomous evaluation and the simulated judgment will be based on the same considerations, considerations about well being,[142] but the particular conception of well being relates to the free choice of goals and relations and is, therefore, an account of well being as determined by willing endorsement.[143] Thus, the argument that the hypothetical claim considered here is based on autonomy rests on a two stage analysis of the individual's desires. Although the parental decision purports to substitute what their child wants, the claim based on autonomy does not crystallize at the point of the parental substituted judgment, it crystallizes upon the subject's autonomous evaluation of the simulated decision.[144]

137 The argument that a legitimate ex post facto evaluation of whether a decision made on one's behalf is the decision one would have made for herself is subject to the argument that one's evaluation could be coloured by hindsight, and that she might have made a different decision in the original position. This caveat applies to all evaluations of decisions and is particularly important in the context of medical decision-making in the face of the failure to disclose medical risks. See, for example, *Sidaway v Bethlem Royal Hospital Governors* [1985] AC 871; *Pearce v United Bristol Healthcare NHS Trust* [1999] 48 BMLR 118; *Chester v Afshar* [2005] 1 AC 134. Legally speaking this is a problematic matter of evidence as it would be in the hypothetical claim. There is not the space here for a detailed consideration of the problems of hindsight in evaluating decisions made on one's behalf.

138 P. Lewis, 'Medical Treatment of Dementia Patients at the End of Life: Can the Law Accommodate the Personal Identity and Welfare Problems?' 13 (2006), *European Journal of Health Law* 219, 219–220.

139 Although she did not exist at the time of the original decision, now she clearly does.

140 From an objective perspective.

141 If, of course, she decides that the parental decision made on her behalf does not reflect her wishes that decision does not then reflect her autonomy and, therefore, her autonomy has not been interfered with.

142 Thanks again to Dimitrios Kyritsis for elucidating this point.

143 J. Raz, *The Morality of Freedom,* New York: Oxford University Press (1988); Chapter 14.

144 The decision to screen out certain embryos could arguably be based on the parent's desires, as opposed to their simulation, of a decision based on the potential child's desires. However, The HFEA focuses on the embryo rather than the potential parents in relation to decisions regarding all the treatments it regulates. The HFEA Code of Practice, eighth edition, makes clear that the likely degree of suffering associated with the condition 10. 6 (b) is crucial in deciding whether PGD is appropriate, showing that the focus is on the potential person and the concern to avoid her potential suffering with a particular disorder, rather than whether the parents want potential persons with a particular disorder screened out.

Not every wrong of this kind will amount to an interference with the interest in autonomy. Although it might be foreseeable that the claimant would rather not have been born with the condition in question,[145] it might be that the particular claimant does not endorse her parent's decision to not select her and is, therefore, grateful for the negligence because it caused her to be born. Here, there is no interference with autonomy which might be caused by way of being born when one would rather not, and could legitimately not, have been born. There is still a legal wrong, but it has not interfered with the created person's autonomy.[146]

The notion of a desire never to have been born is a difficult concept. However, this difficulty arises because of the ability to screen out possible persons on the basis of particular genetic conditions. It is perfectly plausible that a person with a debilitating condition who should not have been born, would, when evaluating the fact of her birth, prefer that she had never been born.[147]

It might be argued that it follows from the claimant's argument that she would rather not have been born that the remedy which she desires is the right to die. However, it is arguable that one can wish that she had never been born without wanting to procure her own premature death. As a result of being born, she may feel that she has formed relationships and responsibilities which prevent her from deciding to end her life, but which would not have been relevant had it not begun.[148] Indeed, it would be unthinkable to mount the argument that in wrongful birth claims the remedy the parents really want is to be able to bring about the death of their child. Even though they might wish the child had never been born, it does not follow that they want to bring about its death.

Rational autonomy

If the courts recognize that there has been a setback to the interest in autonomy of one who is born as a result of negligence and wishes she had not been born, it follows that this claim would potentially be open to any person who should have been screened out. Furthermore, if the hypothetical claim discussed here is recognized in English law, it would be difficult to reject

---

145   From the objective position that this is a condition which is associated with a degree of suffering which is sufficiently serious for the HFEA to permit potential people with this condition to be screened out at the embryonic stage.

146   However, there may be an interference with the parents' autonomy here. See the claim in Chapter 5.

147   Indeed, David Benatar makes a convincing argument that coming into existence is always a harm. Even if the law is not prepared to recognize that life per se might amount to harm where it is wrongfully caused, it might be difficult for the law to maintain that the individual has not suffered a harm where she herself endorses, autonomously and emphatically, the decision to de-select her.

148   For further discussion on the distinction between wanting to die and not wanting to have been born, see D. Benatar, *Better Never to have Been: The Harm of Coming into Existence,* Oxford: Oxford University Press (2006).

similar claims on behalf of people who would have been aborted if there had not been negligence in failing to inform the parents about a potential foetal abnormality. Section 1 (1) (d) of the Abortion Act 1967 has been increasingly liberally interpreted so that foetuses with conditions, which are not generally associated with physical pain and are fairly easily remedied, can be aborted under the Act.[149]

Embryo testing is permitted where there is a significant risk that the person will have or develop a serious physical or mental disability, a serious illness or any other serious medical condition. The HFEA is also adopting an increasingly liberal position concerning the disorders which might be screened out under Schedule 2, s. 1ZA (2) (b) of the HFE Act 1990. This creates the potential for people to be born negligently, and therefore bring a claim arguing that they would prefer not to have been born when they are suffering from a condition which is not generally perceived to be associated with significant physical pain and suffering.[150] This is not a concern where the conception of autonomy adopted is a liberal conception which is devoid of any aspect of rationality. If the regulatory body permits de-selection for less serious conditions, on this conception of autonomy, it would follow that autonomy is interfered with in any endorsed decision to de-select a particular embryo which is carelessly selected.

The HFEA allows parents to de-select those embryos possessing the BRCA 1 genetic mutation. Women who possess this gene are around five times more likely to develop breast or ovarian cancer than those who do not possess the genetic mutation.[151] Thus, in this case, if a claim is brought before any cancer occurs, the individual's suffering might relate to the fact that she lives in the constant knowledge that there is a significant chance that she will suffer cancer at some point, rather than hitherto and current

---

149 For a discussion of the breadth of s. 1(1) (d) of the Abortion Act, see R. Scott, 'Interpreting the Disability Ground of the Abortion Act' 64 (2005), *Cambridge Law Journal* 388. See Joanna Jepson's legal challenge to the late abortion of a 28-week-old foetus for cleft palate in 2001: *Jepson v The Chief Constable of West Mercia Police Constabulary* [2003] EWHC 3318. This liberal position diminishes the argument that the parental decision to abort is one which simulates the autonomous judgment of the future individual because it seems to be focused on the *parent's* desire not to have a child at all, or the *parent's* desire not to have a child with a particular disorder. However, this caveat does not apply in equal force to a claim based on the embryo screening provisions in Schedule 2, s. 1ZA of the HFE Act, which are not concerned with parental desire not to have a child per se and focus on the embryo rather than the parents in the context of specific, named disabilities.

150 Of course it is difficult to objectively perceive the suffering of those who have a condition which the perceiver does not suffer. However, from a comparative respective, the suffering associated with some conditions is evidently greater than others, and where a legal claim is based on a claim that suffering is so bad the individual wishes she had never been born, we might expect the individual's level of suffering to be particularly bad.

151 National Cancer Institute. *SEER Cancer Statistics Review*, 1975–2005. Available HTTP: http://seer.cancer.gov/csr/1975_2005/index.html (accessed 8 February 2010).

physical suffering and limitation.[152] Furthermore, some of the conditions for which the HFEA will license PGD may not significantly interfere with the individual's ability to live a normal life. PKU is on the central list and this can now be managed with little or no side effects by eating a low-phenylalanine diet and taking protein supplements.[153] However, an adult sufferer of cystic fibrosis, polycystic kidney disease, spastic paraplegia or Becker's muscular dystrophy, for example, might have endured significant physical limitation, pain and treatment.[154] On the face of it, any claim based on an authentic wish not to have been born, where one could have legitimately been screened out, is as valid as any other.[155]

152   This type of scenario could arise with respect to any of the late onset conditions for which PGD is licensed, such as familial adenomatous polyposis coli, early onset Alzheimer's disease and Huntington's chorea. See the central list of conditions. Available HTTP www.hfea.gov.uk/pgd-screening.html (accessed 11 May 2010).

153   Available HTTP: http://en.wikipedia.org/wiki/Phenylketonuria (accessed 8 February 2010). This is not to suggest that some sufferers of PKU do not suffer significantly or that there is not a range of suffering associated with particular disorders, but merely that the argument that one would have preferred, in line with her parent's decision, not to have been born, is less likely to evoke the sympathy of the judiciary. Indeed, the law is particularly sensitive to degrees of suffering. The perceived level of suffering is central to the question of whether embryo testing will be allowed in the first place. See Schedule 2, s. 1ZA (2) (b) and Part 10.6 of the Human Fertilisation and Embryology Authority Code of Practice, eighth edition. Furthermore, the courts have not been afraid to draw what might be perceived to be arbitrary limits on the level with regard to particular types of harm, i.e. the distinction between recognized psychiatric harm and grief with respect to harm to the mind.

154   This is, of course, not to suggest that all cystic fibrosis sufferers will feel that their life is not worth living. It is only those who genuinely wish that they had never been born who would have any grievance. Uncomplicated spastic paraplegia is not associated with mental disability. The condition causes progressive spasticity in the lower limbs, leading, inter alia, to difficult walking, ultimately possibly requiring a wheelchair or bed confinement and controlling bladder function. The average age of onset is 24 years, but can be in infancy. See A. K. Erichsen, J. Koht, A. Stray-Pedersen, A. Abdelnoor and C. M. Tallaksen, 'Prevalence of Hereditary Ataxia and Spastic Paraplegia in Southeast Norway: A Population Based Study' 132 (2009), *Brain* 1577. Polycystic kidney disease (PKD) causes enlarged kidneys which can lead to abdominal discomfort, requiring invasive cyst drainage for relief, cyst rupture and peritonitis. PKD can ultimately cause the kidneys to fail. In this case, life preservation (which, of course, might not be what the hypothetical claimant wants) requires dialysis or transplant. Available HTTP: www.bbc.co.uk/health/physical_health/conditions/in_depth/kidneys/polycystickidney1.shtml (accessed 4 June 2010). Becker's muscular dystrophy is characterized by a slow progressive muscle weakness. Eventually the sufferer may become unable to walk and may suffer painful muscle contractions and abnormal bone development, fatigue, breathing difficulties and heart disease. Cognitive function is not usually impaired. Symptoms usually appear between the ages of eight and 25. Available at HTTP: http://en.wikipedia.org/wiki/Becker's_muscular_dystrophy (accessed 4 June 2010). The purpose of the very short evaluation of these conditions is to demonstrate the potential for hypothetical claims based on the created person's autonomous endorsement of her parent's decision that she not be born.

155   Interestingly, the Royal College of Obstetricians and Gynaecologists has released a working party report recommending that a definitive list of conditions that constitute 'serious handicap' for clinicians to use in interpreting the legal grounds for abortion under the 1967 Abortion Act should not be produced because of the kind of difficulties discussed here which might be encountered in predicting and classifying the seriousness of particular conditions. Termination of Pregnancy for Fetal Abnormality in England, Scotland and Wales RCOG Working Party Report. Available HTTP: www.rcog.org.uk/termination-pregnancy-fetal-abnormality-england-scotland-and-wales (accessed 9 July 2010).

One of the reasons the Court of Appeal was keen not to allow the claim in *McKay* was because of a fear that it would mean that a doctor would be obliged to pay damages to a child infected with rubella before birth, who was in fact born with some mercifully trivial abnormality.[156] The courts did not want to reach a situation where compensation would be available for trivial abnormalities, and therefore denied that there was a harm because they would have had little ability to control the boundaries of that harm. Thus, even though the court might recognize that the severely disabled claimant has suffered a deleterious event by being born when she endorses her parent's decision that she not be born, they might reject her claim on the basis that there would be no grounds upon which the courts could distinguish the same kind of claim on behalf of one who suffers a less severe genetic condition.

As argued above, the liberal conception of autonomy which is based on procedural independence alone[157] does not lend itself to external assessment of what autonomy consists in,[158] which might provide analytical purchase for a court seeking to limit the hypothetical claim to those who they perceive to suffer deeply. If the courts adopted a concept of rational autonomy, a negligent failure to prevent the claimant's life would only interfere with the claimant's autonomy where her wish not to have been born is rational.[159]

Relying on rationality does not obviate the argument that the harm is based on the interference with her interest in autonomy. The claim still originates in the claimant's wishes. Where she is pleased she was born, her birth does not interfere with her autonomy. Thus, the catalyst for the harm in each case is still the individual claimant's wish, but that wish can only be legally recognized as interfering with the protected interest in rational autonomy if it meets the condition of rationality. Any approach based on autonomy would require the courts to verify the genuineness of the claimant's wish, which forms the basis of her harm. This is part of the reason why the courts require a verifiable medical condition before they will recognize the harm in cases of harm to the mind. Nevertheless, verifying genuineness is central to the judicial role and where a case for the recognition of claims based on important interests can be made, the refusal to recognize any such claims because of difficulty in verifying genuineness should not lead to a failure to protect those interests. In such cases, it may not be difficult to verify the genuineness of the claimant's wish not to have been born. She might have had a life blighted by

---

156 *McKay v Essex Area Health Authority* [1982] QB 1166, Stephenson LJ, 1181.

157 Different potential legal conceptions of autonomy are discussed in depth in Chapter 3. For supporters of this liberal interpretation of autonomy see, for example, J. P. Plamenatz, *Consent, Freedom and Political Obligation* London: Oxford University Press (1938), 110; T. V. Daveney, 'Wanting' 11 (1961), *Philosophical Quarterly* 139.

158 Over and above establishment of said procedural independence.

159 For an in depth discussion of rational autonomy, see Chapter 3.

treatment, suffering and limitation during which she has not been able to experience things that many people feel bring them happiness and meaning.[160] In these circumstances, the genuineness of the claimant's wishes might be easily verifiable on account of her own evidence and the testimony of witnesses such as family, medical practitioners and others who suffer from that condition but whose lives were not a result of negligence.

If the interest in autonomy which forms a basis for recognizing damage in negligence is rational autonomy, and the interpretation of rationality adopted is based on value or substance rather than procedure,[161] then, as argued in Chapter 3, the court's perception of the ordinary man could be relied on to pour content into the concept of value or substantive rationality.[162] If the values of the ordinary man were relied on to pour content into the issue of a rational desire not to have been born, it is likely that a distinction would arise based on the severity and suffering associated with the condition. Some of the conditions found on the HFEA's central list of conditions, which can be deselected using PGD, are associated with significant and lifelong suffering which presents in childhood. On the other hand, minimal treatment and life changes enable sufferers of some of the conditions to lead a relatively normal life.[163] The courts might perceive that ordinary people would see a wish never to have been born on behalf of a seriously disabled individual to be rational, whilst perceiving the same wish on behalf of a person who, upon minimal treatment, lives a relatively normal life not to be rational. If the courts perceive the quality of life associated with a particular condition to be so bad that rational people would consider the claimant's desire never to have existed with that genetic condition to be rational, then that individual might be deemed to be expressing an autonomous wish never to have existed and, therefore, by causing her to exist against her wishes, the defendant interfered with her autonomy. However, where the level of suffering associated with a particular condition is perceived by the court to be regarded as mild by the majority of ordinary people, they might consider a wish not to have been born not to be rational and, therefore, not rationally autonomous.

Some might object to this approach on the basis that it requires the court to make distinctions between disabilities from a perspective of not having experienced that disability. However, the view that it would be better that the claimant had not been born originates in the claimant. It is not a

---

160   Examples might be: getting an education or career, engaging in sports or hobbies, forming relationships and having children. Not all people want the same thing, but most people do not aspire to spend much of their life in hospital, having treatment.

161   See Chapter 3 for a discussion of these two interpretations of what rationality consists in.

162   In Chapter 3 it was argued that one way that the courts might pour content into the notion of rationality would be to adopt the English tort system's intrinsic standard for attributing value to potentially protectable novel interests; the standard of the ordinary person.

163   See the discussion above.

judgment being made about those who are incapable of forming a view of the desirability of their own births, or those who suffer the same fate as the claimant but consider themselves lucky. The concept of autonomy recognizes that people can have different views and reactions to their disability.[164] This view is the catalyst for the perspective that the birth was an interference with autonomy, but the question of whether there is a legal harm requires an element of objectivity. As we know, the Californian courts recognize that an appeal might be made to the perspective of the societal consensus with regard to the distinctions which might be made regarding the types of condition which might merit recognition of harm in wrongful life actions.[165]

Nevertheless, it remains difficult for any court to make value rationality judgments based simply on the severity of a disability, whatever the level of societal consensus. Thus, it might be argued that rather than ascribing value in terms of rationality to the claimant's wish per se, the courts could adopt a substantive interpretation of rationality which focuses on the reasons for the claimant's desire.[166] From this perspective, the claimant would be required to demonstrate to the court's satisfaction that she has reasonable reasons for wishing never to have been born, if that wish is to be deemed rational and, therefore, autonomous. Morgan and Veitch consider the need for individuals who appear to have odd desires and decisions to be able to convince the court that their reasons for the decision are somehow valid,[167] relying on the case of Ms B[168] to highlight the fact that when a decision seems odd in abstract, that puts the claimant in the position of having to justify that decision. In the context of the autonomy-based genomic negligence claim considered here, the claimant will be required to speak of her desire never to have been born and the suffering that she has experienced through being born. Through this, she will have to convince the court that she has rational reasons for her desire.[169]

---

164 Compare Daniel James' reaction to becoming paralyzed after a rugby scrum collapsed on him with that of Hillary Lister who is paralyzed due to a progressive illness. Daniel James travelled to Switzerland with his parents to lawfully end his life at the Dignitas clinic in 2008. On 31 August 2009 Hillary Lister became the first paraplegic woman to sail solo around Britain.

165 *Turpin v Sortini* (1982) 643 P 2d 954, Kaus J, 963. See the court's discussion of the distinction between the severity of the condition in *Curlender* and *Turpin* in *Turpin*. Although rationality here relates to the claim based on the disability itself, as opposed to the claimant's autonomous evaluation of the life with the disability. See also the discussion of the Californian cases of *Turpin* and *Curlender* earlier in this chapter.

166 This alternative approach to value rationality was considered in detail in an abstract sense in Chapter 3.

167 D. Morgan and K. Veitch, 26 (2004) 'Being Ms B: B, Autonomy and the Nature of Legal Regulation', *Sydney Law Review* 107.

168 *Re B (Consent to Treatment: Capacity)* [2002] 1 FLR 1090.

169 Rational reasons are thus, determined in much the same way as the determination of the substantive rationality of the actual wish discussed above; that is, the court's perception of what the reasonable person would deem rational.

This approach does not require that the courts see value in the desire itself in abstract; instead, they would need to be able to see value in the reasons for that desire that would lead them to empathize with the claimant so as to see why, in the light of those reasons, an apparently odd desire might be rational and therefore autonomous.

Relying on Thomas Nagal's paper, 'What is it like to be a bat?', Kim Atkins argues that all experience is ascribable to a subject to count as an experience, and we cannot isolate our understanding of that experience from the perspective of the subject whose experience it is.[170] Given the nature of this experience, it is impossible to know objectively what it would be like to be another human being. However, Atkins concludes that humans' bodies and persons are sufficiently similar for us to at least imagine roughly the subjective character of certain situations for certain people if our life experiences and values are sufficiently similar. For Atkins, recognizing the subjective character of experience is at the heart of empathy and allows us to reflect genuine respect for the individual's autonomy. She argues that it is to these intimations that we must appeal if we value autonomy once we admit that autonomy is expressive of what it is like to be that person.[171] Requiring individuals to give reasons for their apparently odd wishes or decisions, allows the decision-assessor to get closer to the subjective experience of the individual, perhaps allowing her to see why although the decision appears odd in abstract, it is rational from the perspective of the individual in the position of the individual.[172] Thus, the individual must convince the court that she has rational reasons for an apparently irrational desire. The introduction of this element of subjectivity might enable the decision-assessor to see as rational that which she might intuitively see as irrational.

Conclusion

The increase in the availability and uptake of embryo screening raises the potential for carelessness to lead to the birth of a person whose birth was not intended. Where that person experiences an existence which is blighted by suffering, and deeply wishes that her parent's decision that she not be born

---

170 K. Atkins, 'Autonomy and the Subjective Character of Experience' 17 (2000), *Journal of Applied Philosophy* 71, 72.

171 Ibid. 78.

172 A similar point is made in the context of IVF treatment whereby rational reasons can be given for an apparently odd decision to undergo treatment which has a tiny chance of success. See L. Koch, 'IVF – An Irrational Choice?' 3 (1990), *Reproductive and Genetic Engineering: Journal of International Feminist Analysis* 235. Koch argues, although IVF seems an irrational choice to feminist critics, because the risks are high and success rates are low, infertile women do not seem to hear the well-argued and well-founded warnings of critical feminists because these women have good reasons to try IVF.

had been carefully respected, should the courts dismiss her claim on the basis that she has suffered no harm? Currently, the position in English law is that those who exist and suffer because of another's wrong in failing to prevent their birth do not suffer harm. However, this approach presupposes a conception of harm which rests on the concept of welfare. It is possible to interpret the harm suffered differently. If the process of decision-making in the hypothetical claim is seen as a two-stage process, consisting of the parents substituted judgment and the subject's autonomous evaluation of that simulated decision as, in fact, what she wanted, it might be argued that the harm to the individual consists of the interference with her autonomy. From this perspective, the wrong occurs when the person who was entrusted with preventing the unintended birth acts without due care and attention, therefore causing that unintended birth, but the interference with autonomy crystallizes upon the unintended person's evaluation and endorsement of the decision to prevent her birth. In these circumstances, it might be argued that the conception of autonomy adopted needs to be further refined if it is to form the basis of legally recognizable harm. A rationality-based conception of autonomy is considered here. However, whatever the conception of autonomy relied on, some legal recognition of the individual's autonomous desires seems appropriate in the context of a clear wrong and a legal system which purports to recognize autonomy as a fundamental principle in medical law,[173] and an increasingly important principle in negligence law.[174]

---

173 For expressions of this view, see R. S. Downie and K. C. Calman, *Healthy Respect: Ethics in Health Care,* second edition, Oxford: Oxford University Press (1994), Chapter 4; R. S. Downie, C. Tanna-hill, A. Tannahill, *Health Promotion: Models and Values,* second edition, Oxford: Oxford University Press (1996), Chapters 9 and 10. For legal expressions of the fundamental nature of autonomy in the medical context, see the discussion in Chapter 3.

174 See the House of Lords decisions in *Rees v Darlington Memorial Hospital NHS Trust* [2004] 1 AC 309 and *Chester v Afshar* [2005] 1 AC 134.

Chapter 5

Negligence in reproductive genetics

The parents' perspective

Introduction

Similarly to the preceding chapter, this chapter considers the grievance that might arise where a careless failure to properly test an embryo leads to the birth of a child, whose birth the parents sought to avoid. However, here the focus is on the claims that parents might bring. The potential claims on behalf of the parents are considered within the context of saviour sibling treatment because this enables the discussion of a range of potential parental grievances. The Human Fertilisation and Embryology Authority (the Authority) issued its first licence to create a saviour sibling in 2001. Despite some strong objections,[1] the Authority has since relaxed the eligibility criteria, increasing the number of people who could access such treatment. The permission to create saviour siblings has now been given statutory authority by virtue of the 2008 amendment to the 1990 Human Fertilisation and Embryology Act.[2] In the face of a rising demand for treatment, it may only be a matter of time before an attempt to create a saviour child goes wrong, generating new grievances that might be articulated as novel legal claims.

There are two reasons why an intended saviour might not be able to fulfil the saviour role. First, the saviour herself might have the faulty gene which caused the existing child's condition. Second, although she does not possess the faulty gene, the saviour might not be a tissue match for her sibling. A multitude of tort actions could (in principle) be brought by the

---

1 The Human Genetics Commission consultation paper, *Choosing the Future: Genetics and Reproductive Decision Making*, (2004) paragraph 3.17, notes that the use of preimplantation genetic diagnosis (to any end) is not uncontroversial. Those who take a pro-choice approach support the procedure on the basis that it facilitates the right to reproductive freedom. See, for example, the judgment of the House of Lords in Quintavalle's challenge of the Human Fertilisation and Embryology Authority's decision to give the go-ahead for the Hashmis: *Quintavalle v Human Fertilisation and Embryology Authority* [2005] 2 AC 56, Lord Hoffmann, 570.

2 Human Fertilisation and Embryology Act 1990, s. 1ZA (1) (d).

children or the parents affected in these scenarios.[3] The discussion is split
into three parts.

By way of background, Part I outlines the history of the creation of
saviour siblings in this jurisdiction. Part II considers the existing law on
the birth of unwanted children. The discussion addresses the fact that in
wrongful conception claims the English courts appear to have abandoned
normal legal principles in favour of an approach which seeks to ensure a
just distribution of burdens in society. This part focuses on how this
approach has led the courts to draw distinctions between children who are
born with disabilities and those who are not, which are not meaningful
from a legal perspective. These rather empty distinctions have particular
relevance to new claims, which might arise from the birth of the wrong
child following treatment to procure a saviour sibling. Of particular interest here, is the claim that might be brought on the basis of the continuation of the child's existing illness which would most likely have been
ameliorated if the created child had, as intended, been able to act as a
saviour. Part III argues that the construction of the harm where the
parents bear a child who is not wanted because of her particular characteristics might be more realistically construed as an interference with autonomy.[4] Such an approach might provide a more comprehensive method of
dealing with these types of claim and may prevent the need to make the
empty distinctions between cases which pervade judicial decisions in this
area. Reformulating the harm in this way would also enable redress which
properly reflects the loss experienced. With respect to the question of what
the parents have lost in terms of autonomy, the focus is on what the value
of autonomy consists in. This leads into a discussion of whether autonomy
has value in the sense that interference with it amounts to the setback to
an interest in and of itself, or whether the value of autonomy is contained
in what autonomy makes possible. Finally, this two-fold value perspective
is applied to the novel hypothetical scenario concerning parents whose
attempt to create a saviour for their sick child has been negligently
frustrated.

---

3 As this chapter focuses on parental claims, it ignores the potential wrongful life claim of the disabled
   saviour child because this is considered in the previous chapter. It also ignores the potential claims
   that might be brought directly by the existing child. An interesting consideration would be whether a
   tort system, which is imbued with recognition of the interest in personal autonomy, would recognize
   that the failure to secure a cure for the existing child's serious illness amounts to a failure to protect
   her potential to maximize her autonomy as a person without serious illness. The loss here is probably
   most catastrophic for the sick older sibling whose life cannot now be saved by the new child's birth, or
   who has to continue with invasive and painful treatment. Thank-you to Professor Emily Jackson for
   elucidating this point. This claim is not considered here, but is the subject of a separate work in
   progress.
4 As was suggested by the majority of the House of Lords in *Rees v Darlington Memorial Hospital NHS
   Trust* [2004] 1 AC 309.

Background

Zain Hashmi has thalassaemia. His only hope of a cure is a stem cell or bone marrow transplant from a tissue-matched donor. In 2001, unable to find a donor, Zain's parents sought preimplantation genetic diagnosis (PGD) to test their embryos to ensure they were (a) free of thalassaemia and (b) a tissue match for Zain, so that the umbilical cord blood of the resulting child could provide life-saving treatment for Zain. Both the Hashmis carry the thalassaemia gene, thus their embryos were eligible for PGD to screen out the disorder. The crux of the issue was whether they could also screen in an embryo with a particular tissue-type. In December 2001 the Human Fertilisation and Embryology Authority announced that it was prepared to allow tissue typing where PGD was already indicated for a severe or life threatening disorder.[5] Legal wrangling[6] and controversial distinctions between similar cases ensued.[7] However, the legitimacy of saviour sibling treatment has now been established by s. 1ZA (1) (d) of the HFE Act 1990, and its availability no longer depends on the embryo being at risk of the disorder. Given the multitude of IVF mix-ups that have occurred in recent years,[8] it might be argued that it is only a matter of time until an attempt to create a saviour sibling is negligently frustrated, leading to the birth of a child who is not as her parents wanted her to be.

If the parents brought a claim in these circumstances, it would rest on the premise that they did not want *this* child. That is; they wanted a child, but not a child with the characteristics that this child has, as opposed to not wanting a child at all. This situation is similar to the situation where carelessness has led to the failure to diagnose a foetal abnormality in a wanted pregnancy, thereby depriving the mother of her right to abortion. These types of claims are popularly referred to as 'wrongful birth' claims. The terms 'wrongful conception' or 'wrongful pregnancy' usually refer to situations where the defendant was negligent in failing to prevent conception per se. In these cases the parents did not want a child at all. The novel claim

---

5   Human Fertilisation and Embryology Authority Press Release, 'HFEA to Allow Tissue Typing in Conjunction with Preimplantation Genetic Diagnosis' (13 December 2001).

6   *Quintavalle v Human Fertilisation and Embryology Authority* [2005] 2 AC 561, Lord Hoffmann, 570.

7   See the grounds upon which the Human Fertilisation and Embryology Authority rejected the application for saviour sibling treatment on behalf of the Whittaker family for their son, Charlie. Human Fertilisation and Embryology Authority Press Release, 'HFEA Confirms that HLA Tissue Typing May Only Take Place when PGD is Required to Avoid a Serious Genetic Disorder' (1 August 2002). See also the Authority's about turn in 2004 in response to an application on behalf of the Fletcher family: Human Fertilisation and Embryology Authority Press Release, 'HFEA Agrees to Extend Policy on Tissue Typing' (21 July 2004).

8   Four separate events occurred at Leeds Teaching Hospitals. B. Toft, *Independent Review of the Circumstances Surrounding Four Adverse Events that Occurred in the Reproductive Medicine Units at the Leeds Teaching Hospitals NHS Trust,* West Yorkshire, Department of Health (2004). The review concluded that the events were caused by a mixture of inadvertent human error and systems failure.

considered here does not fit squarely into the ambit of wrongful conception of wrongful birth. The timing of the carelessness occurred preconceptually, so there appears to be something wrongful in the conception itself. Nevertheless, the woman *did* want to conceive so the loss does not lie in the conception per se. The loss becomes apparent at, or soon after, birth and relates to the existence of the particular child rather than a child per se.

Although some commentators maintain that there is a distinction between these two types of claim, others have criticized these categorical labels:

> These labels are not instructive. Any 'wrongfulness' lies not in the life, the birth, the conception, or the pregnancy, but in the negligence of the physician. The harm, if any, is not the birth itself but the effect of the defendant's negligence on the parents' physical, emotional and financial well-being resulting from the denial to the parents of their right, as the case may be, to decide whether to bear a child or whether to bear a child with a genetic or other defect.[9]

The judgments in English wrongful birth cases draw heavily on the cases concerning wrongful conception. Nevertheless, it is clear that whilst the courts have refused in recent years to award damages for the direct consequences of the negligence in wrongful conception cases,[10] they have not been so reluctant to award damages for the consequences in wrongful birth cases.[11] Indeed, the position in relation to wrongful birth appears settled, whilst the legal position with respect to wrongful conception is generally acknowledged to be in a mess.[12]

Although at first glance it might seem odd, the novel claims discussed here will largely be considered in the context of the existing law on wrongful conception rather than wrongful birth. There are three major reasons for this. First, the hypothetical parents' claim might be more similar to wrongful conception claims than straightforward wrongful birth claims. Although many parents seeking saviour sibling treatment want another child in any event, there is at least a suspicion that they might feel differently if they did not have an ill child who could be cured by a tissue matched sibling.[13] Thus, it might be argued that these parents did not primarily want another child,

---

9 *Viccaro v Milunsky* 551 NE 2d 8 (Mass 1990) 9, n. 3.

10 Since *McFarlane v Tayside Health Board* [2000] 2 AC 59.

11 See, for example, *Rand v East Dorset* Health Authority [2001] PIQR Q1; *Hardman v Amin* [2000] Lloyd's Med Rep 498; *Taunton v Somerset NHS Trust* [2000] All ER (D) 2460.

12 M. Lunney, 'A Right Old Mess: *Rees v Darlington Health Authority* [2003] 3 WLR 1091' 1 (2004), *University of New England Law Journal* 154.

13 This was the position of parents seeking to obtain a license from the HFEA prior to the 2008 amendments to the 1990 Human Fertilisation and Embryology Act. However, saviour sibling treatment is now permitted by statute: s. 1ZA (1) (d), HFEA 1990, and there is no requirement that the prospective parents demonstrate that they would have wanted another child in any event.

primarily they wanted a treatment for their existing child and the child is incidental to this.[14] If it was not possible to procure a saviour sibling these parents might not have sought to have any more children. In this way, these parents are different to the parents in wrongful birth claims who want a child per se. The hypothetical parents' desire to have a child is based on the desire to obtain the outcome that is frustrated by the negligence. In most wrongful birth claims the desire to have a child is independent of the medical intervention, which the negligence claim stems from, and is an outcome wanted in and of itself; that is, the desire itself is not linked to the desire to attain something which the negligence frustrates.[15]

Second, the focus here is not simply on recovery for the costs of the created child's disability.[16] However, the wrongful birth cases are explored when the particular formulation of the novel claims based on the existing child's disability is considered below. The focus here is on the particularly novel formulation of the hypothetical claim which might be put when the saviour child is healthy, but not the saviour she was expected to be.[17] Here the argument is that the loss might be tied to the disability of others. The issue of others' disabilities has been considered within the context of wrongful conception,[18] but not wrongful birth, thereby providing a basis from which to consider the novel legal issue of the responsibility for the continuation of the existing child's disability.

Third, the latter part of this chapter argues that the negligence has occasioned loss to the parents on the basis that it has interfered with their autonomy. Given this, it seems more relevant to focus on the decisions of the Court of Appeal in *Parkinson v St James and Seacroft University Hospital NHS Trust* and the House of Lords in *Rees v Darlington Memorial Hospital NHS Trust* where the prospect that the loss concerned the interference with autonomy, as opposed to the parents' economic interests.[19] Nevertheless, this is not to say that reformulating the damage as an interference with autonomy is not relevant in wrongful birth cases. Here, as in wrongful conception, it might be argued that the real loss caused by the fact that the child is not as she was intended to be, lies in the interference with the parents' wishes and the subsequent implications of this interference upon their life plan, rather than solely in the interference with their finances.

---

14 I do not intend this to be an objection to saviour sibling treatment. My argument here is not to suggest that the saviour child does not become a much loved member of the family.

15 It might be possible to contest this distinction where there is a prior known risk of disability in cases of wrongful birth, and this was the reason the parents sought medical assistance.

16 If she has the very disability that the parents were seeking to avoid in her and cure in their existing child via her.

17 Because she does not possess the 'right' tissue type; that is, the type matching the existing sick child's tissue.

18 On this, see *Rees v Darlington Memorial Hospital NHS Trust* [2004] 1 AC 309.

19 The wrongful birth cases have concentrated solely on the extraordinary economic costs associated with the child's disability.

Wrongful conception

The healthy child – McFarlane v Tayside Health Board

Three fairly recent claims shed light on how the English courts might react to parental claims arising when there is a careless failure in screening potential saviour embryos. Mr and Mrs McFarlane claimed the costs of raising an unwanted child whom they conceived after receiving negligent advice that Mr McFarlane had been rendered infertile after a vasectomy.[20] The House of Lords acknowledged that the pregnancy and the birth of the child were direct and foreseeable consequences of the incorrect information, and that ordinary tortious principles pointed towards recovery.[21] Nevertheless, the House chose to view the case from the perspective of distributive justice rather than that of corrective justice.[22]

According to Lord Steyn, the concept of distributive justice required the court to strive to achieve a just distribution of burdens and losses among members of a society.[23] He thought this might be achieved by asking the commuter on the London Underground whether the parents of an unplanned, but healthy, child should be able to sue the doctor or hospital for compensation equivalent to the cost of bringing up the child. Lord Steyn was of the opinion that an overwhelming number of ordinary men and women would answer this question with an emphatic 'No', based on their view as to what is morally acceptable and what is not. He was supported on this point by Lord Hope and Lord Millett.[24] Thus, although ordinary legal principles pointed towards recovery, notions of distributive justice led the House to conclude that allowing parents to recover the costs of raising a healthy child ran counter to the values which they held, and which they believed that society at large could be expected to hold. *McFarlane* represented an about turn in English law, which had, for the previous fifteen years, recognized that where a child would not have been conceived and born 'but for' negligent treatment, damages could be awarded for the costs of child rearing.[25]

20 *McFarlane v Tayside Health Board* [2000] 2 AC 59.
21 Ibid. Lord Steyn, 82; Lord Hope, 95; Lord Millett, 107. See also *Rees v Darlington Memorial Hospital NHS Trust* [2004] 1 AC 309, Lord Nicholls, 318; Lord Hutton 338; *Parkinson v St James and Seacroft University Hospital NHS Trust* [2002] QB 266, Hale LJ, 288; *Cattanach v Melchior* [2003] HCA 38, Kirby J, 158.
22 *McFarlane v Tayside Health Board* [2000] 2 AC 59, Lord Steyn, 82.
23 Ibid.
24 Ibid. Lord Hope, 97; Lord Millett, 113–114.
25 *Emeh v Kensington, Chelsea and Westminster Area Health Authority* [1985] QB 1012; *Thake v Maurice* [1985] 2 WLR 215; *Bennarr v Kettering HA* (1988) 138 NLJ 179; *Gold v Haringey Health Authority* [1988] QB 481; *Allen v Bloomsbury HA* [1993] 1 All ER 651; *Crouchman v Burke, The Times* (10 October 1997); *Allan v Greater Glasgow Health Board*, 1998 SLT 580.

One of the problems with seeking a remedy for the costs of raising a child in negligence is that the birth of the child does not fall within the traditional conceptions of harm, which the tort recognizes, because tort law is never happier than when faced with a knotty problem involving a collision between strangers, preferably with lots of broken limbs.[26]

In *McFarlane*, the Lords categorized the harm in question as pure economic loss.[27] The characterization of the loss as purely economic creates internal inconsistencies. On the one hand, the House allowed Mrs McFarlane's claim for the pain and inconvenience of the pregnancy and for those expenses arising during the pregnancy,[28] on the basis that those economic losses were a consequence of the personal injury of pregnancy.[29] On the other hand, they rejected both parents' claims for the child rearing costs on the basis that, unlike the costs arising during the pregnancy, such costs were purely economic,[30] the implication being that such costs were not a direct consequence of the pregnancy. Priaulx asks: can maintenance costs be correctly characterized as a 'pure economic loss?'[31] If pregnancy is a personal injury, the economic loss suffered by the mother is immediately consequential on that injury. She concludes that their Lordships did actually recognize the harm in *McFarlane,* but nevertheless declined to provide a remedy.[32]

Others have criticized the decision to prevent, on policy grounds, what would, on ordinary negligence principles, be recoverable.[33] One critic responded with particularly strong condemnation:

I can think of few decisions that are to their very core as odious, unsound and unsafe as this one.[34] The Australian High Court also found little to *McFarlane* and refused to follow it.[35]

Nonetheless, it now appears to be settled law that the cost of raising an unplanned healthy child is not recognizable loss in English law. However, the English courts have since heard two factual variants of *McFarlane*: *Parkinson v Seacroft University Hospital NHS Trust*;[36] and *Rees v Darlington Memorial Hospital*

---

26 J. Conaghan, 'Tort Law and Feminist Critique' in M. D. A. Freeman (ed.), *Current Legal Problems,* Oxford: Oxford University Press (2004), 174, 192.

27 *McFarlane v Tayside Health Board* [2000] 2 AC 59, Lord Slynn, 76; Lord Steyn, 79; Lord Hope, 89; Lord Clyde, 100.

28 Ibid. Lord Clyde, 98; Lord Slynn, 74.

29 Ibid. Lord Steyn, 84; Lord Hope, 86–87; Lord Clyde, 102.

30 Ibid. Lord Steyn, 79.

31 N. Priaulx, 'Joy to the World! A (Healthy) Child is Born! Reconceptualizing "Harm" in Wrongful Conception' 13 (2004), *Social & Legal* Studies 5, 13.

32 Ibid. 13.

33 A. Pedain, 'Unconventional Justice in the House of Lords' 63 (2004), *Cambridge Law Journal* 19, 19.

34 J. E. Cameron-Perry, 'Return of the Burden of the Blessing' 149 (1999), *New Law Journal* 1887, 1888.

35 *Cattanach v Melchior* [2003] HCA 38.

36 *Parkinson v Seacroft University Hospital NHS Trust* [2002] QB 266.

Trust.[37] Although the courts did not challenge the *McFarlane* decision, they made exceptions to it, which do not appear to be based on any material legal distinction.

The disabled child – Parkinson v Seacroft University Hospital NHS Trust

When she gave birth to a disabled son after a negligent sterilization, Mrs Parkinson sought to recover the costs of raising him. The Court of Appeal acknowledged that conventional tort principles pointed towards full recovery,[38] but bound by *McFarlane*, declined to award the ordinary costs of raising the child. Nevertheless, the court held that *McFarlane* did not preclude an award of compensation for the extra-ordinary costs of raising a disabled child.[39]

Similarly to Lord Steyn in *McFarlane*, Brooke LJ sought to approach the case on the basis of the views of ordinary people as opposed to strict legal principle.[40] He felt that ordinary people *would* consider it fair for the law to award damages if they were limited to the extra expenses associated with the child's disability.[41] Hale LJ also distinguished the case from *McFarlane*, arguing that, although ordinary people would consider it fair to refuse the McFarlanes' claim, they would not consider it unfair if the person who had undertaken to prevent conception, pregnancy and birth negligently failed to do so, was held responsible for the extra costs of bringing up a disabled child.[42]

The problem with *Parkinson* is that the defendant's negligence primarily caused the birth, rather than the disability. There was no known risk of disability; thus, in all material legal respects, the case was identical to *McFarlane*. In both cases, the harm had arisen from the birth of a child, and in both cases this was what the defendant had been employed to prevent. The creation of this arguably arbitrary distinction set the stage for a claim to be brought for extra costs associated with a disability suffered by someone other than the wrongfully conceived child.

---

37 *Rees v Darlington Memorial Hospital NHS Trust* [2004] 1 AC 309.
38 *Parkinson v Seacroft University Hospital NHS Trust* [2002] QB 266, Hale LJ, 288–289.
39 In *Parkinson*, Brooke LJ, 283, proceeded on the basis that the loss was economic loss, but he considered this to be consequential rather than pure. Hale LJ, 286, however, did not perceive the loss as wholly economic. She thought pregnancy involved a severe curtailment of one's personal autonomy, which did not stop once the mother had returned to her pre-pregnancy state, because of her parental responsibility for the child.
40 It is possible that this approach falls within the scope of legal principle because of the considerable latitude offered by the just, fair and reasonable label. I owe this point to Dr Nicolette Priaulx.
41 *Parkinson v Seacroft University Hospital NHS Trust* [2002] QB 266, Brooke LJ, 283.
42 Ibid. Hale LJ, 295.

The disabled mother – Rees v Darlington Memorial Hospital NHS Trust

Karina Rees underwent sterilization because she felt her visual impairment would make it difficult to care for a child. However, she conceived her healthy son after the operation was performed negligently. She claimed damages for the costs of raising him. In the House of Lords, she sought to uphold the Court of Appeal's decision to award the extra costs attributable to her disability, but also claimed the whole cost of raising her child, inviting the House to reconsider *McFarlane*.

Again, on an orthodox application of conventional tortious principles, the Lords conceded that the full costs of rearing the child should be awarded.[43] Nevertheless, all seven held that the *McFarlane* decision to categorize the ordinary costs of child rearing as non-recoverable pure economic loss should stand. Despite this, none of them thought the disabled mother should be sent away empty handed. Their Lordships were split 4:3 in their view as to what the appropriate remedy was. The minority agreed with the Court of Appeal that the mother should be awarded the extra costs of raising the child incurred by her disability.[44] However, the majority felt that the minority's decision would lead to the drawing of indefensible distinctions in an effort to create a principled line between costs which are recoverable, and those which are not, which would render the law incoherent.[45] According to Lord Millett:

> ordinary people would think it unfair that a disabled person should recover the costs of looking after a healthy child when a person not suffering from a disability who through no fault of her own was no better able to look after a child could not.[46]

The majority thought that the proper outcome was to award a conventional sum, not to compensate for the birth of the child, but simply to recognize the wrong done.[47] If damages are based on the wrong done rather than the loss in terms of the costs of raising a child conceived 'wrongfully', it follows that the conventional sum should be applicable in all cases where an unplanned child is born following a negligent sterilization.[48] As in *McFarlane* and *Parkinson*, the

---

43 See, for example, *Rees v Darlington Memorial Hospital NHS Trust* [2004] 1 AC 309, Lord Bingham, 314; Lord Nicholls, 318; Lord Millett, 344.

44 *Rees v Darlington Memorial Hospital NHS Trust* [2004] 1 AC 309, Lord Steyn, 327.

45 Ibid. Lord Millett, 348–349.

46 Ibid. Lord Millett, 349.

47 *Rees v Darlington Memorial Hospital NHS Trust* [2004] 1 AC 309, Lord Bingham, 316–317; Lord Nicholls, 319; Lord Millett, 349; Lord Scott 356. The minority was critical of the conventional sum. See Lord Steyn, 328; Lord Hope, 335.

48 The speeches in *Rees* support this. See, for example, Lord Bingham 316–317; Lord Nicholls, 319; Lord Millett, 349.

reasoning of neither the majority nor the minority in *Rees* was based on conventional principles of English negligence law. After this trilogy of cases, it might be argued that the search for legal principle in this area has been abandoned.[49]

Policy concerns: a question of priorities

The English appellate courts have adopted an approach whereby their response to wrongful conception claims is referenced to the particular judge's perception of the ordinary person's likely response to the claim. Although legal principles pointed towards recovery for the costs of raising each of these children, the essence of the loss led the courts to conclude that the claims were of low priority, thus ordinary tort principles were not adhered to. There are many reasons why the courts might want to prioritize claims. First, we know from constant reiteration that the courts are under pressure and are therefore keen to limit the number of potential claims.[50] Where the courts believe they would be unable to cope with potential liability they prioritize those claims they believe to be the most deserving.[51] In a similar vein, the threat of a significant number of claims against a defendant whom the courts might wish to shield from an increase in liability might encourage the prioritization of claims. In the context of wrongful conception, the primary defendant is likely to be the National Health Service (NHS).[52] Given that the NHS is publically funded, an entity which provides a service of significant utility, and whose constant financial difficulties are well documented, it is not surprising that the courts appear eager to protect the NHS from liability.[53]

Where established legal principles are limited on the basis of priorities there should be a clear basis as to what gives a claim priority. In the context of wrongful conception claims the courts have judged priorities by reference to the values of the ordinary person. In Chapter 3 it was argued that whilst

---

49 Lord Hoffmann took the view that the search for principle had been called off in the field of psychiatric harm when he relied on the principle of distributive justice to reject the police officers' claims in *White v Chief Constable of South Yorkshire* [1999] 2 AC 455. See also *Parkinson v Seacroft University Hospital NHS Trust* [2002] QB 266, Brooke LJ, 274.

50 See, for example, *Ultramares Corporation v Touche* (1931) 255 N.Y. 170, Cardozo CJ,179; *Alcock v Chief Constable of South Yorkshire Police* [1992] 1 AC 310, Lord Oliver, 417; *White v Chief Constable of South Yorkshire Police* [1999] 2 AC 455, Lord Griffiths, 464; Lord Steyn 492.

51 The problem with this approach lies in determining which cases are the most deserving. For an account of how the courts have determined this in the context of wrongful conception claims, see discussion below.

52 *Bionews*, 'NHS to Fund Saviour Sibling Treatment' (22 March 2004). Available HTTP: www.bionews.org.uk/page_11894.asp (accessed 16 June 2010).

53 See, for example, *Rees v Darlington Memorial Hospital NHS Trust* [2004] 1 AC 309, Lord Bingham, 316; Lord Nicholls, 319 and Kirby J's interpretation of the reasons for the decision in *McFarlane* in *Cattanach v Melchior* [2003] HCA 38, 178. See also M. A. Jones, 'Another Lost Opportunity' 19 (2003), Professional *Negligence* 559, who suggests that *Wilsher v Essex AHA* is an anomaly in the field of causation because of the fact that the defendant in that case was the NHS.

taking into account the views of society in the formulation of legal policy has inherent logic,[54] in the absence of empirical evidence of the views of ordinary people, the judge's perception of those views might not be accurate. There may be clear-cut scenarios where one option is so hideous that it is reasonable to conclude that the majority of ordinary citizens would come to the same conclusion. However, the wrongful conception cases discussed here would not evoke such a clear and unanimous reaction. Indeed, Lord Steyn acknowledged this in *Rees* when he said: 'The issue in *McFarlane* was a profoundly controversial one. Ultimately, there was a choice to be made between eminently reasonable competing arguments.'[55]

It is difficult to see then how Lord Steyn could have been so sure in *McFarlane* that an 'overwhelming number of ordinary men and women' would choose one of two eminently reasonable options.[56] Perhaps the decision to accord low priority to wrongful conception claims is motivated by the categorization of the loss as purely economic. Presenting the harm as purely financial ignores the realities of raising a child and fails to recognize that what one loses when one has a child is much more than financial.[57] Whilst many people might not be moved by an unwilling parents' complaint that the child costs money, they might be moved where the complaint relates to the fact that the unwanted child necessitates a complete change in the life plan that the parents had set their hearts on. Framed in this way, it is arguably not clear that the ordinary person would think justice could only be served by denying the parents' claim. At the very least, on this construction of the harm, if the question were put to the commuter on the Underground it would be unlikely that an 'overwhelming majority' would be opposed to the parents' claim.[58]

Raising a child is a particularly life-changing experience. For the majority, especially given the availability of abortion and contraception, rearing a child is a happy event. However, we should not assume that because this is largely the case, it ought always to be the case.[59] Although child rearing

---

54 C. Witting, 'Physical Damage in Negligence' 61 (2002), *Cambridge Law Journal* 189, 194.

55 *Rees v Darlington Memorial Hospital NHS Trust* [2004] 1 AC 309, Lord Steyn, 324.

56 Or why the judges in *Parkinson* were of the opinion that the majority of ordinary people would consider that *McFarlane* had been correctly decided. See *Parkinson v Seacroft University Hospital NHS Trust* [2002] QB 266, Brooke LJ, 283; Hale LJ, 295.

57 Of course there can also be many benefits to having a child, but this hinges on whether that child is wanted.

58 See Lord Steyn's formulation of the question and answer of the hypothetical commuter in *McFarlane v Tayside Health Board* [2000] 2 AC 59, Lord Steyn, 82.

59 Consider the case of a 16 year old student whose significantly older boyfriend has a vasectomy, having already had children of his own. If the vasectomy is negligently performed with the result that the teenager becomes pregnant, would the courts hold that the birth of the child should be seen by the parents as a joy and a blessing? In the context of the political agenda to reduce teenage pregnancy they might see that the failure to prevent *this* pregnancy and birth has significant adverse consequences on the mother's life plan.

involves significant financial output, an unwanted[60] child might be more damaging to a person's life plans than her finances. Career-focused individuals who like nothing more than to indulge in adventurous pursuits and holidays in their free time might not feel aggrieved by the money they spend raising an unwanted child, but they might harbour significant grievance about the lifestyle which child-rearing requires them to leave behind. This is not to suggest that an unwanted child is only a non-financial hardship for wealthy, adventurous people. Peoples' reasons for not wanting a/ another child will be varied. Some people might not want children because of disability or the existence of other dependants. Others might simply want to lead a life unburdened by the responsibility of a/another child. From this perspective, it might be argued that the real damage that is suffered in the face of wrongful conception is the loss of the ability to live one's life as one wishes to. This acknowledges the real and lasting effect that an unwanted child might have, in that it might force a person to adopt a wholly different life-plan to the one which she had firmly set her heart on. This does not change the fact that the only remedy available in this scenario is financial.[61] The courts cannot alter the fact that the child now exists.[62]

Many might think the interference with deep rooted life plans is significantly more damaging than the economic loss that rearing a child entails. If the loss were framed in these terms, it may be more difficult for the court to conclude that the majority of ordinary people would think the parents' claim to be of insufficient priority for compensation. Where the courts own test for ascertaining whether a claim is of sufficient priority for compensation is unable to yield a result for a claim that would on ordinary tortious principle be recoverable, it might be argued that the benefit of any doubt should lie with the claimant.

Within the context of this discussion of the courts policy of ordering priorities for compensation, let us throw another factual variant into the melting pot: a couple, seeking a saviour sibling, bring a novel wrongful

---

60 Some commentators have struggled with the pejorative nature of the term 'unwanted child' and therefore sought to apply a different label. In *Richardson v LRC Products Ltd* [2000] Lloyds Med Rep 28, Kennedy J suggested the term uncovenented child for this reason. However, I have, like some other commentators, chosen to stick with the term 'unwanted'. Concerns about the term have generally related to the problem of children finding out that they were not wanted. However, the term 'unwanted' seems to most accurately reflect the parents' reaction to the situation at the time whether they subsequently love and care for the child or not.

61 An argument that a financial remedy is not appropriate where the damage is not financial could not be mounted without challenging the system of recovery for tortious injury generally.

62 The parents could have pursued abortion or adoption. But the House of Lords was adamant that the McFarlanes' failure to arrange an adoption or an abortion was not a *novus actus* which broke the chain of causation. *McFarlane v Tayside Health Board* [2000] 2 AC 59, Lord Slynn, 74; Lord Clyde, 104; Lord Millett 113.

conception claim, when, due to negligence on behalf of the medical team, their intended saviour a) is healthy, but her tissue does not match that of her sibling or, b) has the very disorder she was supposed to cure in her sibling.

The hypothetical claim: the healthy child – low priority?

Although since *McFarlane* the English courts have refused to interpret the harm in wrongful conception cases as loss consequent upon physical injury,[63] writing extra-judicially, Lady Justice Hale (as she then was) has expressed support for this position: 'First, left to myself, I would not regard the upbringing of a child as pure economic loss, but loss which is consequential upon invasion of bodily integrity.'[64]

The problem with framing the hypothetical wrongful conception claim as consequential economic loss is that the consequential approach rests on the premise that the pregnancy itself is personal injury. This categorization of the loss is based on the woman's desire not to achieve a pregnancy, and therefore to regard the pregnancy as a deleterious occurrence. The problem for our hypothetical parents is that the mother intended and wanted to be pregnant and the discovery of her pregnancy was most likely a joyous occasion. Furthermore, the parents are likely to face further problems in convincing a court that they should be compensated for the costs of raising the child in any event. The parents wanted to have *a* child and intended and expected to incur the costs of raising that child. Thus, they have not lost anything in terms of the costs of child rearing. However, it might be argued that the fact that the parents in the hypothetical claim analysed here wanted to have *a* child,[65] serves to distinguish the claim from wrongful conception actions. The hypothetical parents' claim does not rest on a rejection of the institution of parenthood per se. Their claim arises because of their desire to have another child and it is based on their commitment to being the best parents they can to their existing child. On the other hand, the wrongful conception claims arise because the claimants did not want to become parents, whether on this particular occasion or at all.[66] Thus, the latter claim rests on a rejection of the institution of parenthood.[67] Any claim which rests on the rejection of parenthood in and of itself is unlikely to be recognized if the law adopts a pro-natalist perspective. A society's prevailing social norms might

---

63 Which is the physical state of pregnancy.
64 Dame Brenda Hale, 'The Value of Life and the Cost of Living – Damages for Wrongful Birth', The Staple Inn Reading (2001), *British Actuarial Journal* 755, 761.
65 Albeit a particular type of child; a saviour one.
66 I am grateful to Dr Nicolette Priaulx for pointing out this distinction in the motivations of the parents in the hypothetical case and the wrongful conception cases.
67 In this specific case or generally. Thanks go again to Dr Nicolette Priaulx for this point.

reflect dominant pro-natalism or anti-natalism. As Fletcher notes, the one child laws in contemporary China show how social pressure to stem population growth in a territory deemed to be overpopulated can take a general anti-natalist form. The constitutional protection of the right to life of the 'unborn' in the Republic of Ireland reflects a general pro-natalist form as only those women whose lives are seriously threatened are legally permitted to terminate a pregnancy.[68] However, Fletcher also notes that it 'is also very common to find pro-natalist and anti-natalist trends within the one legal system'.[69] In this jurisdiction the law allows the sterilization of incapacitated women[70] where it is deemed to be in their best interests.[71] Although it cannot be in a person's best interests to be sterilized for the purposes of eugenic or population control reasons,[72] the lawfulness of sterilization does not rest on there being therapeutic reasons for the procedure.[73] Discussing the legal sanctioning of the sterilization of 'mentally deficient'[74] individuals in Canada and the USA, Fletcher argues that 'the desire is not for fewer children per se, but for fewer of particular *kinds* of children'.[75] Thus, a society can have some anti-natalist policies within a largely pro-natalist ethic.

Whilst, at first glance, the Abortion Act 1976 appears to be a piece of legislation which is underpinned by an anti-natalist ethic because the Act only permits particular types of abortion, it combines pro-natalist and anti-natalist ideology. The UK abortion legislation takes the position that a decision to abort is not in and of itself an acceptable one.[76] It is not a decision which a woman can make simply because she does not want a child; she has to fulfil specific criteria which mean that only those women who have good reasons, such as the wrong type of foetus or poor social conditions, will be allowed to terminate a pregnancy.[77] The UK legislation is relatively speaking lenient. However, this does not interfere with the argument that the 1967 Act does not accord a right to women that would allow them to seek abortion simply because this is what they desire. They need to prove that theirs is the type of case in which the anti-natalist ethic should be allowed to

---

68 R. Fletcher, 'Legal Forms and Reproductive Norms' 12 (2003), *Social and Legal Studies* 217, 232.

69 Ibid. 232

70 The Court of Appeal refused to allow sterilization of a man on the basis of best interests: *Re A* [2000] 1 FLR 549.

71 See, in particular, *Re F* [1990] 2 AC 1.

72 *Re B* [1988] AC 199.

73 *Re F* [1990] 2 AC 1. However, see *Re S* [2001] Fam 15 for the view that surgical sterilization should be the option of last resort even where there is some therapeutic reason why surgical sterilization might be indicated.

74 R. Fletcher, 'Legal Forms and Reproductive Norms' 12 (2003), *Social and Legal Studies* 217, 232.

75 Ibid. (original emphasis)

76 S. Sheldon, *Beyond Control, Medical Power and Abortion Law*, London: Pluto Press (1997), 42.

77 Ibid.

override the pro-natalist ethic. On this basis, Sheldon argues that abortion stands as the 'exception to the norm of maternity'.[78]

It might be argued that the wrongful conception cases discussed here are underpinned by a pro-natalist ethic.[79] Indeed, a recurring theme throughout these cases is that the judges felt that it was unacceptable for the parents to view their child as a harm. In *McFarlane*, Lord Steyn felt that 'the law of tort had no business to provide legal remedies consequent upon the birth of a healthy child, which all of us regard as a valuable and good thing'.[80] Lord Millett expressed similar sentiments, finding that 'the law must take the birth of a normal, healthy baby to be a blessing, not a detriment'.[81] In *Rees v Darlington Memorial Hospital NHS Trust* all five members of the House maintained the *McFarlane* result, in effect refusing to challenge the view that the birth of a healthy child should be seen as a good and valuable thing. In particular, Lord Millett remained true to the view he put forward in *McFarlane*: parenting a healthy child is a valuable experience which the law should not recognize as a harm.[82]

Thus, it might be argued that the essence of the wrongful conception cases is a refusal to recognize claims which can be seen as rejecting the valuable institution of parenthood. However, the hypothetical claim is not based on the rejection of parenthood.[83] The claimants wanted to become parents again. In this way the ethic underpinning their reproductive project is pro-natalist and would, therefore, fit with the House of Lords' view that we should all value the birth of healthy children. Furthermore, the hypothetical parents want to be good parents to their existing child by doing all they can to relieve her suffering and prolong her life.[84] Thus, their claim is not problematically anti-natalist in the same way as the wrongful conception claims.

If the lack of an anti-natalist basis makes the judiciary feel positively inclined to the hypothetical claimants, then the decisions in *Parkinson* and *Rees* present two possibilities for compensation. *Parkinson* presents the possibility

---

78 Ibid.

79 I owe this point to Dr Nicolette Priaulx.

80 *McFarlane v Tayside Health Board* [2000] 2 AC 59, Lord Steyn, 82.

81 Ibid. Lord Millett 114–115. Similarly, Lord Hope, 97, said

> There are benefits in this arrangement as well as costs. In the short term there is the pleasure which a child gives in return for the love and care which she receives during infancy. In the longer term there is the mutual relationship of support and affection which will continue well beyond the ending of the period of her childhood.

82 *Rees v Darlington Memorial Hospital NHS Trust* [2004] 1 AC 309, Lord Millett, 346.

83 I owe this point to Dr Nicolette Priaulx.

84 Perhaps the kind of parents who would not be allowed to access abortion under the 1967 Act (except for the reason of foetal abnormality) because they are the kind of parents to whom pro-natalist policy applies.

of claiming for the costs associated with the *existing child's* or the *created child's* disability, and *Rees* presents the possibility of claiming a conventional sum in either case. We can deal with the issue of recovering the costs of the disabilities first.

The costs of the existing child's disability

The costs of the unwanted child's disability were awarded in *Parkinson*. Furthermore, a significant minority of the House and majority of the Court of Appeal would have awarded the costs associated with the mother's disability in *Rees*. Therefore, it is possible that, if faced with a *Rees*-type case, a differently constituted House might re-examine the conventional sum and award the costs associated with the mother's disability. In the face of this uncertainty, let us consider how the courts might react if the costs of another party's disabilities were claimed; namely the costs associated with an existing child's disability where her intended saviour cannot cure her.

Generally, wrongful conception claims argue that the negligence led to the birth of an unwanted child and that negligence can be said to have caused foreseeable loss associated with raising that child. However, where the claim is for the costs of an unwanted child's incidental disability, the same direct link between the negligence and the harm cannot be established. Mrs Parkinson did not seek sterilization because she was at an elevated risk of bearing a disabled child; the negligence related to the birth of the child, not his disability. Similarly, in *Rees*, the mother's disability was not caused by the defendant's negligence. In both cases we know that some level of compensation was awarded because of the existence of the disability, albeit more directly in *Parkinson*. In *Rees*, the majority did not consider Ms Rees' disability to be crucial to the award of the conventional sum. However, when this result is considered in the light of *McFarlane*, which their Lordships refused to interfere with in *Rees*, the position remains that the disabled parent was afforded a measure of compensation which was denied to the able parent.[85] Whether this will translate into a concrete legal distinction between disabled and able bodied parents remains to be seen. Although the minority in *Rees* would have based their award on the mother's disability, the disability was not crucial to the reasoning of the majority. Thus, until *McFarlane* mark two, it is not clear whether able bodied parents will receive the conventional sum.[86]

If the *Rees'* conventional sum is available in all wrongful conception cases, it is not clear whether the *Parkinson* exception, where the extraordinary costs

85 Furthermore, a substantial minority in *Rees* was against the conventional sum and would have afforded damages based on the mother's disability.
86 I am grateful to Professor Emily Jackson for pointing out this position.

of the unwanted child's incidental disability can be recovered, will be maintained. The distinctions drawn in the trilogy of cases discussed here are arguably based on legally immaterial facts, resulting in tough to justify distinctions and making it difficult to predict judicial reaction to the novel claim.

Nevertheless, the causal caveat, which creates problems in *Parkinson* and the minority reasoning in *Rees,* would not apply in a claim for the extra costs associated with the disability of an existing child in the frustrated attempt to create a saviour. The continuation of the existing child's disability relates directly to the medical team's failure to provide the intended treatment.[87] However, in *Rees,* the majority was critical of the decision to base compensation on the disability. Lord Bingham said:

> I would for my part apply this rule (the conventional sum) also, without differentiation, to cases in which either the child or the parent is (or claims to be) disabled. While I have every sympathy with the Court of Appeal's view that Mrs Parkinson should be compensated, it is arguably anomalous that the defendant's liability should be related to a disability which the defendant's negligence did not cause and not to the birth which it did.[88]

Similar sentiments were expressed by the rest of the majority who stressed that anomalies would arise if a distinction were made on the basis of disability.[89] Given the House of Lord's reluctance to create anomalies by distinguishing between wrongful conception cases based on the involvement of a disabled party, the possibility of an award based on the existing child's disability looks dubious even in the face of the more direct causal connection.

The costs of the created child's disability

Assume now that lack of due care and attention on behalf of the medical team leads to the saviour child having the very disorder that the parents were seeking to avoid. Again, the parents clearly wanted to have a child, making it difficult for them to maintain a claim for the costs of raising the child per se. The question here is whether they ought to be able to claim for the extra costs associated with the saviour child's disability.

Presume that the saviour child is not the saviour she was expected to be because she suffers from the same disorder as the existing child. The parents might claim that although they wanted another child, they did not want

---

87 A stem cell transplant between tissue matched siblings carries a greater than 90 per cent success rate. Communication with Professor Ajay Vora, professor of paediatric haematology and consultant paediatric haematologist (Professor Vora treated Charlie Whitaker after his parents' project to create a saviour sibling for him in Chicago was successful).

88 *Rees v Darlington Memorial Hospital NHS Trust* [2004] 1 AC 309, Lord Bingham, 317.

89 Ibid. Lord Nicholls, 319; Lord Millett, 346.

another child with the illness that they were trying to cure in their existing child. At first glance it seems that the hypothetical parents should be awarded the extra costs of the created child's disability when the extra costs of the child's disability were awarded in *Parkinson.* It is well established that if the parents claim is a wrongful birth claim, which seeks only the extra costs associated with the child's disability, they would be able to recover those costs.[90] In the light of *Parkinson,* there is now similar scope for recovery of these extra-ordinary costs when the claim is for wrongful conception. Nevertheless, we know that the majority in *Rees* was critical of the decision in *Parkinson* to compensate for the extra costs of the child's disability. In the face of this criticism, it is not clear how a future House of Lords will decide a *Parkinson*-type case. However, despite the similarities between *Parkinson* and the hypothetical case, there are also significant differences that might persuade the courts that the hypothetical is of higher priority for compensation than a *Parkinson*-type claim.

Although some judges might find the decision to award compensation in a *Parkinson*-type claim dubious on the basis that there is an insufficient causal link between the negligence and the harm,[91] where the parents specifically employ the defendant to select an embryo without a particular disability, the court might conclude that there is a sufficient causal link with the created child's disability. Although in *Rees* Lord Scott was critical of the decision in *Parkinson,* he suggested that there should be a distinction between cases where the avoidance of the birth of a child with a disability is the very reason why the parents sought treatment to avoid conception, and cases where the medical treatment was sought simply to avoid conception.[92] Although this reasoning is based on foreseeable disability arising from negligent sterilization, it seems that Lord Scott would limit compensation to the extra-ordinary costs of the unwanted child's disability. On this basis, there is no reason why the hypothetical claim should not fit squarely within this exception to the no recovery rule.

However, it might be argued that an award of the extra-ordinary costs of the saviour child's disability does little to reflect the actual loss. The financial costs associated with the child's disability are likely to largely be met through social welfare in the form of Disability Living Allowance[93] and NHS

---

90 *Rand v East Dorset Health Authority* [2001] PIQR Q1; *Hardman v Amin* [2000] Lloyd's Med Rep 498; *Lee v Taunton and Somerset NHS Trust* [2001] 1 FLR 419.

91 Compensation was awarded on the basis that it was foreseeable that a disabled child would be born where the defendant was negligent in preventing the birth because of the one in 200 to one in 400 chance of congenital abnormality in any live birth. See, for example, *Parkinson v Seacroft University Hospital NHS Trust* [2002] QB 266, Brooke LJ, 272. See also *Rees v Darlington Memorial Hospital NHS Trust* [2004] 1 AC 309, Lord Scott, 356.

92 *Rees v Darlington Memorial Hospital NHS Trust* [2004] 1 AC 309, Lord Scott, 355.

93 This is available for disabled people to meet the extra cost of living expenses relating to their disability. Available HTTP: www.direct.gov.uk/disabledpeople/financialsupport (accessed 14 April 2010).

medical care. Moreover, the extra financial outlay associated with the child's disability might not be particularly significant.[94] On the other hand, the parents might perceive that the birth of another child with a serious disability has considerable emotional and time implications and will impinge significantly on their life plans. If this is the case, they might want to claim an award that they believe more accurately reflects the loss suffered. The argument that the loss in these reproductive claims might be refocused to better reflect the actual loss was recognized by the House of Lords' fairly recent decision in *Rees* to award a conventional sum for the birth of an unwanted child.

A conventional sum?

The focus for recovery of the conventional sum in *Rees* was the wrong, as opposed to the harm.[95] Therefore, it should be no obstacle that the hypothetical parents may have wanted a child in any event and, therefore, would have incurred the costs of raising *a* child even if it was not *this* child. Traditionally, damage has formed the gist of the action in negligence.[96] Thus, if their Lordships were in fact making an award simply on the basis that there was a wrong in *Rees*, this amounts to a particularly radical development in the law of negligence.

So let us analyse more deeply what the conventional sum purports to represent. The judgments in *Rees* suggest that the purpose of the conventional sum was to recognize that the interference with the parents' autonomy was a legal wrong which the law should recognize. According to Lord Millett, the purpose of the conventional sum was to recognize the denial of an important aspect of their autonomy, viz the right to limit the size of their family'.[97] Although they did not expressly state that the award was designed to recognize the breach of Ms Rees' autonomy, Lord Bingham[98] and Lord Nicholls[99] expressed similar sentiments.

---

94   Particularly if the condition in question is one which involves intensive medical treatment which is provided by the NHS, as opposed to significant living adaptations for physical or mental disability.

95   *Rees v Darlington Memorial Hospital NHS Trust* [2004] 1 AC 309, Lord Bingham, 316–317; Lord Millett, 349; Lord Scott, 356. However, some academics argue that the conventional sum is actually recognition of a new form of damage, as opposed to the simple recognition of a wrong. See D. Nolan, 'New Forms of Damage in Negligence' 70 (2007), *Modern Law Review* 59; N. Priaulx, *The Harm Paradox: Tort Law and the Unwanted Child in an Era of Choice,* Oxford: Routledge-Cavendish (2007).

96   J. Stapleton, 'The Gist of Negligence. Part I: Minimum Actionable Damage' 104 (1988), *Law Quarterly Review* 213.

97   *Rees v Darlington Memorial Hospital NHS Trust* [2004] 1 AC 309, Lord Millett, 349.

98   Ibid. Lord Bingham, 317.

99   Ibid. Lord Nicholls, 319.

Although breach of autonomy is well recognized in the context of consent to medical treatment, such that damages for trespass to the person are available where health professionals deliberately refuse to fulfil the wishes of a competent adult,[100] the failure to respect one's wishes is not something which the law of negligence generally recognizes. Negligence is focused on damage, and damage has a fairly narrow meaning; usually harm to body or property, and in tightly circumscribed circumstances, mental and financial loss. The values protected by autonomy are more intangible, but they relate to respecting people by treating them as autonomous and recognizing that other values, such as contentment and self-fulfilment, may be reliant on the exercise of autonomy.[101] When the value of autonomy is framed in these terms, it is clear that the conventional sum bore little resemblance to the interference with autonomy which was recognized in *Rees*. Comprehensive recognition of an interest in autonomy might involve a more realistic assessment and reflection of the impact that a negligent failure to respect one's reproductive wishes can have on the individual.

A new approach based on interference with autonomy

The value of autonomy

Many academic commentators believe that the true loss where a person's reproductive choices are interfered with is one of autonomy.[102] According to Jackson, overarching regulation of the reproductive sphere should be concerned with promoting autonomous decision-making.[103] Priaulx comprehensively analyses the adoption of the concept of autonomy as a means of protecting reproductive choice, specifically in the context of wrongful conception.[104] She argues that when reproductive choices are 'defeated

---

100 See, in particular, *Re B (Consent to Treatment: Capacity)* [2002] 1 FLR 1090.

101 This perspective demonstrates the link between the values of autonomy and well being. For more on this link, see the discussion in Chapter 3 and J. Raz, *The Morality of Freedom*, New York: Oxford University Press (1988), Chapter 14.

102 N. Priaulx, 'Joy to the World! A (Healthy) Child is Born! Reconceptualizing 'Harm' in Wrongful Conception' 13 (2004), *Social & Legal* Studies 5; N. Priaulx, *The Harm Paradox: Tort Law and the Unwanted Child in an Era of Choice*, Oxford: Routledge-Cavendish (2007); J. Conaghan, 'Tort Law and Feminist Critique' in *Current Legal Problems* M. D. A. Freeman (ed.), Oxford University Press (2004); E. Jackson, *Regulating Reproduction: Law, Technology and Autonomy*, Oxford: Hart (2001). For a discussion of reproductive autonomy in the context of advances in genetic reproductive technology see Human Genetics Commission, 'Making Babies: Reproductive Decisions and Genetic Technologies' (January 2006), paragraph 1.3.

103 E. Jackson, *Regulating Reproduction: Law, Technology and Autonomy*, Oxford: Hart (2001), 1.

104 N. Priaulx, *The Harm Paradox: Tort Law and the Unwanted Child in an Era of Choice*, Oxford: Routledge-Cavendish (2007); N. Priaulx, 'Joy to the World! A (Healthy) Child is Born! Reconceptualizing "Harm" in Wrongful Conception' 13 (2004), *Social & Legal* Studies 5, 6–7.

through negligence the parents must confront a different life plan, one that holds inescapable parenting obligations, including financial, social and psychological implications', such that the principle of autonomy most obviously arises as an 'interest capable of being defeated through unsolicited parenthood'.[105]

Autonomy is a matter of self-authorship which requires choices, actions and desires to be respected. In theory, it might be argued that all tortious interferences involve some infringement of the victim's autonomy because given the choice, most people would not have chosen to be mentally or physically injured by the defendant's negligence.[106] The damages awarded in negligence for such injuries will reflect the lasting interference that the injury has on the claimant's life. The purpose of tortious damages is to put the victim in the position she was in before the tort was committed. Of course, with respect to physical losses, particularly to the person, a return to the status quo may be impossible, but the basic principle is that of full and adequate compensation. Some go as far as saying that the English tort system prides itself on the provision of such complete compensation.[107]

The principle of full compensation provides for the recognition of the ways in which negligence has interfered with the particular victim's life plan. Damages in negligence are earnings related to, rather than based on, flat rates. In this way, damages are tailored to the individual's previous circumstances and how those circumstances might have continued if she had not been injured. The tort of negligence is aptly able to compensate vague and difficult-to-quantify-financially losses which are based on the victim's inability to enjoy those things that she enjoyed before the tort was committed. On the basis of loss of amenity, losses such as impairment of one's sexual life,[108] an inability to go fishing[109] and an inability to play with one's children,[110] are compensated.

The ability of damages in negligence to reflect negligent interferences with one's individual life plans depends on the existence of legally recognizable damage in the first place. The problem with the reproductive claims is that the House of Lords has declared that the birth of a child is not recognizable damage. However, following *Rees*, the House of Lords has effectively recognized the loss that occurs when there is an interference with personal autonomy, but has gone against tortious damages principles in failing to

---

105   Ibid. 15.
106   R. Mullender, 'English Negligence Law as a Human Practice' 21 (2009), *Law and Literature* 321, 324–325.
107   S. Deakin, A. Johnston and B. Markesinis, *Tort Law*, fifth edition, Oxford: Oxford University Press (2003), 792.
108   *Cook v JL Kier & Co Ltd* [1970] 2 All ER 513.
109   *Moeliker v Reyrolle & Co Ltd* [1977] 1 All ER 9.
110   *Hoffman v Sofaer* [1982] WLR 1350.

award complete compensation for that loss. This contradicts the fundamental principle that the purpose of tortious damages is to put the victim in the position he was in before the tort was committed because it makes no assessment of the impact that the tort has had on the claimant's position. Furthermore, the award of a conventional sum, which is based on a flat rate as opposed to an assessment of the harm to the particular claimant, does not sit well with the aim to provide individually tailored compensation. In essence, although the conventional sum purports to protect autonomy, in reality it does little to reflect the protection of autonomy which rests on individual choices, actions and desires and the effects that interference with autonomy can have on one's life plans. The question of how a particular event interferes with an individual's autonomy is not one that can be met with a clear and uniform answer. It will depend on individual factors; namely, what were the individual's plans? And to what extent has the negligence curtailed them?

If autonomy is to be recognized as an interest worthy of legal protection then, as Priaulx argues, it is essential that we articulate what it is about autonomy that is valuable.[111] Priaulx argues that our understanding of the worth of autonomy must extend beyond the notion of mere choice.[112] If autonomy is ever to hold any moral *value* or potential in our lives at all, then it must be because our choices are guided towards the aim of human flourishing or towards the living of a good life.[113] If the courts are to recognize an interest in personal autonomy, they should consider whether the value of autonomy consists of the product of one's choice, action or desire as a reflection of her preferences, or in being able to make choices and have desires independent of the effects that might be produced. From the perspective of the reproductive claims considered here, the question of where the value lies in autonomy becomes particularly important in deciding what loss the claimant has suffered. It was argued in Chapter 3 that a true recognition of the value of autonomy requires us to recognize that autonomy is an intrinsic good. However, although recognition of the intrinsic value of autonomy is a necessary condition in protecting the essence of what it is that is valuable about autonomy,[114] it might be argued that it is not sufficient where the effects of the interference with autonomy extend to the frustration of the things which the exercise of autonomy sought to make possible.

---

111 N. Priaulx, *The Harm Paradox: Tort Law and the Unwanted Child in an Era of Choice*, Oxford: Routledge-Cavendish (2007), 184.

112 Ibid.

113 Ibid. (original emphasis). Latter part citing A. McCall Smith 'Beyond Autonomy' 14 (1997) *Journal of Contemporary Health Law and Policy* 30. See also J. Raz, *The Morality of Freedom*, New York: Oxford University Press (1988), Chapter 14 for a discussion of the link between autonomy and well being. This relates to the discussion of what is valuable about autonomy in Chapter 3. There the issue is framed in terms of intrinsic and instrumental value.

114 That is, that autonomy is something that is worth experiencing for its own sake.

It is a good thing in and of itself that we are able to experience forming our own desires and make choices and actions based on these desires, but the making of our own choices and actions is also a good thing as a means to what it makes possible: the fulfilment of personal goals and contentment. The whole value of autonomy is a sum of its instrumental and intrinsic parts. Despite this, it is suggested in Chapter 3 that if the English courts were to recognize an interest in personal autonomy within the tort system, on the basis of practical concerns, they would recognize the instrumental value of autonomy rather than the intrinsic value or whole value. The basis for this argument is that tangible loss is more easily perceived where interference with a person's choice has an effect on her future plans. Despite this, it seems that in *Rees* the House of Lords unwittingly recognized the intrinsic value of autonomy, but not the instrumental or whole value. In *Rees*, Lords Bingham and Nicholls were keen to point out that the purpose of the conventional sum was to recognize the wrong done rather than the consequences of that wrong.[115] Coupled with the fact that the sum was an arbitrarily set amount, it is clear that the award was not designed to reflect the de facto impact upon Ms Rees' life plans that were caused by the failure to respect her desires and corresponding choices. On the face of it, it seems as though the sum was based simply on the fact that her choice was interfered with. However, it remains to be seen whether this is a correct interpretation of the court's intention because the essence of this interpretation is that the value of autonomy lies in the experience of having one's desires and choices respected *per se*. In theory, this would mean that a claimant who found out that her desire to be sterilized had been frustrated before it led to any relevant consequences, namely before a pregnancy occurred, would also be entitled to recover this sum. It is, however, doubtful whether the House of Lords would see such a claim as of sufficient priority for compensation. However, a protection of autonomy does not recognize the full value if the principle does not recognize that the intrinsic value of autonomy is interfered with in the event of an interference with choice *per se*, only recognizing the intrinsic value where there has, in fact, been some interference with instrumental value. That is, in *Rees*, even though the damages were specifically not designed to compensate for the birth of the child,[116] the consequences of the negligent interference with the choice were crucial to the recognition of the interference with autonomy.[117] Thus, somewhat inimically in *Rees*, the House of

---

115 *Rees v Darlington Memorial Hospital NHS Trust* [2004] 1 AC 309, Lord Bingham, 316–317; Lord Nicholls, 319.

116 Ibid. 317; Lord Millett, 350, held that the conventional sum was not designed to compensate for the birth of the child.

117 I.e. where the failure to respect the person's choice does not impact upon her life plans.

Lords recognized only the intrinsic value of autonomy,[118] but in circumstances where the interference with the person's choice has had some
further impact upon the aim which the exercise of autonomy sought to
achieve.

Is there a difference between deliberate and negligent breach of autonomy?

Legally speaking, it seems that the value of autonomy relates to the manner in
which the interference with autonomy occurs. Whilst the deliberate interference with a person's decision regarding medical treatment has long been
recognized within the English tort system, the notion that careless interference
with medical decision-making might be a focus for legal recognition is only
just emerging.[119] The distinction between deliberate and careless interference
with choices and desires might shed some light on whether the value of autonomy in the legal context is considered to be intrinsic or instrumental.

It is a fundamental principle of English medical law that an adult patient
who suffers from no mental incapacity has an absolute right to choose
whether to consent to medical treatment, to refuse it or to choose one rather
than another of the treatments being offered.[120] Where a person capable of
autonomous choice is deliberately treated against her will, the tort of
battery will provide a remedy irrespective of whether the interference with
that choice has some further deleterious impact on the things which the
individual sought to achieve via the choice. The tort is actionable per se,
and seeks to protect against unwanted interference as opposed to only those
interferences that cause further damage. In this way, the tort of battery recognizes that having one's choices and desires respected is valuable in and of
itself. Assume that an individual has an objection to taking a particular pill
perhaps because she does not want to subject herself to the risks associated
with taking that pill.[121] However, the health professional feels it is in the
patient's best interest to take the pill and, therefore, she gives it to her
along with her other pills without telling her that it is the unwanted pill.
The patient suffers no ill effects from taking this pill,[122] but, in time, she
discovers the health professional's failure to respect her choice and feels

---

118 I.e. and not also the instrumental value.
119 This sort of recognition came about in *Rees v Darlington Memorial Hospital NHS Trust* [2004] 1 AC
     309 and *Chester v Afshar* [2005] 1 AC 134.
120 For example, *Re T (Adult: Refusal of Treatment)* [1993] Fam 95, Lord Donaldson, 102; *Re MB (An
     Adult: Medical Treatment)* [1997] 2 FLR 426, Butler-Sloss LJ, 432.
121 It is assumed here that those risks are properly disclosed by the health professional so this is not to
     be confused with a case of non-disclosure of risk.
122 Maybe it even makes her feel better.

significantly aggrieved. For the patient, there was value in the experience of having her choice respected by the health professional. The sense of dignity and self-respect that one experiences when her legitimate choices are respected may be significantly damaged where there is a deliberate failure to respect those choices. In the event of a battery action, the courts might have significant sympathy with the patient in the face of the health professional's deliberate failure to respect the intrinsic value of the patient's autonomy. Nevertheless, battery is also capable of recognizing when adverse consequences *do* flow from an interference and compensating in full for those consequences. Given this, battery is capable of recognizing the whole value of autonomy.

This discussion demonstrates the difference between the level of protection afforded to autonomy in the medical context where there is a deliberate failure to comply with the patient's choice, and the level afforded where the interference with autonomous choice is careless. Might the House of Lords have been prepared to reflect the whole value of Ms Rees' autonomy if the interference with her choice had been deliberate? It is difficult to imagine how such circumstances might occur, but rogue doctors do exist.[123] Consider, for example, how the courts might react if a doctor or pharmacist with a pro-life conscience chose to give a woman a placebo as opposed to the morning after pill. Might the courts be more prepared to recognize the whole value of autonomy rather than the limited value they recognized in *Rees*?

It might be that the culpable party's intention with regard to the wrong ought to lead to the disparate recognition of what is valuable about autonomy. That is, a deliberate disregard for an individual's autonomous choice or desire is somehow worse than a careless disregard, and this should be reflected in the interpretation of what it is of value that the claimant has lost. Intrinsically speaking, it might be more damaging for people to know that their autonomy was deliberately, rather than carelessly, interfered with, but this cannot be assumed. Instrumentally speaking though, the effects of the failure to respect the individual's choice are the same irrespective of whether the interference with the choice was deliberate or careless. Moreover, it might be argued that a thoughtful, deliberate disregard based on good

---

123   A particular relevant example of rogue medical practice occurred in Connecticut when Dr Ben Ramaley allegedly inseminated one of his patients with his own sperm as opposed to that of her husband, leading to the birth of twin girls. The woman's husband was subsequently shown by way of DNA analysis not to be the biological father of the girls. The couple sued Ramaley, alleging that he had 'intentionally inserted his own sperm' into the wife causing her to become pregnant. The case was quickly settled out of court and a gag order was imposed. Ramaley willingly forfeited his New York medical license because of the case.

faith[124] is no more culpable than a careless disregard, which arises from practice which is bad because of a thoughtless lack of respect for the purposes of individuals.

Chester v Afshar[125] adds another perspective to the question of where the value lies in the legal protection of autonomy. Miss Chester suffered partial paralysis after undergoing surgery on her spine. It was established that this injury was an inherent risk of the treatment and the surgeon had not performed the operation negligently. However, the surgeon had been negligent in failing to inform Miss Chester that the operation carried a 1–2 per cent risk of the particular paralysis. Miss Chester admitted in her evidence that whilst if she had been informed of the risk she would not have undergone the operation at that time, she would probably have had it at some time in the future. Thus, Miss Chester could not fulfil the ordinary causal principles.[126] Despite this, the majority held that she should be compensated on the basis that the law should protect the patient's right to make informed choices about medical treatment.[127] Lord Steyn said: 'A rule requiring a doctor to abstain from performing an operation without the informed consent of a patient ... ensures that due respect is given to the autonomy and dignity of each patient.'[128] But what is it about informed consent that protects autonomy? Does the value derive solely from the opportunity to accept those risks that one is willing to accept and obviate those risks that one is not willing to accept? If the value of autonomy rests on being able to avoid risk, recognition of interference with autonomy requires materialization of this risk. From this perspective there will be no interference with the claimant's desire to obviate a risk if, fortuitously, that risk does not eventuate. This reflects an instrumental valuation of what it is about autonomy that informed consent protects. On the other hand, it might be argued that what is valuable about informed consent in terms of protecting autonomy is

---

124   As has occurred in many medical law cases where doctors have disregarded what they perceive to be unwise decisions. See, for example, *Re B (Consent to Treatment: Capacity)* [2002] 1 FLR 1090; *Re MB (An Adult: Medical Treatment)* [1997] 2 FLR 426; *St George's Healthcare NHS Trust v S, R v Collins Ex p. S (No. 2)* [1999] Fam 26; and many cases concerning mature minors. For example *Re E (A Minor) (Wardship: Medical Treatment)* [1993] 1 FLR 386; *Re L (Medical Treatment: Gillick Competence)* [1998] 2 FLR 810; *Re M (Medical Treatment: Consent)* [1999] 2 FLR 1097. Many of these decisions were interpreted as concerning the issue of capacity, but in reality they rest on the fact that the decision made by the patient was considered to be medically unwise. For comment on this issue see, for example, I. Kennedy, 'Commentary on Re MB' (1997) 5, *Medical Law Review* 317, 323; S. Sheldon & M. Thomson, *Feminist Perspectives On Health Care Law*, London: Cavendish Publishing (1998), 243.

125   *Chester v Afshar* [2005] 1 AC 134.

126   Because she could not demonstrate that 'but for' the defendant's negligence she would not have suffered the harm, or that the negligence made a material contribution to her harm.

127   *Chester v Afshar* [2005] 1 AC 134, Lord Hope, 162; Lord Walker, 164; Lord Steyn, 144.

128   Ibid. Lord Steyn, 144.

respecting the individual as someone who is entitled to the relevant information needed to form desires and make choices. There is value in being treated as a person who is able to understand and be able to cope with information which is relevant to making a personal choice about medical treatment. One might feel that it is an affront to her self-determination if relevant information is withheld on the paternalistic grounds that she could not cope with it/ understand it, or it might produce a decision which the person obtaining informed consent deems unwise. Furthermore, it might be argued that the sense of grievance is not limited to intentional failures to disclose information, but is just as likely to arise where the failure is careless or reckless. Here the value of autonomy is intrinsic and lies in being respected as being able to be the author of one's own life and desires, and to make one's own choices accordingly, and this requires relevant information. It might be that one cannot know if she would have chosen differently if she was given the relevant information, but this is irrelevant from an intrinsic perspective. What is crucial is that the individual would have felt that the causal responsibility for her choice lies with her.

Miss Chester's award was directed at the actual costs of the paralysis she suffered. Thus, the value of autonomy was determined by the value of the object which the exercise of autonomy sought to achieve; the ability to obviate risks, rather than the inherent respect in being treated as capable of receiving and understanding information and making a meaningful choice. Despite the fact that the value of autonomy was instrumentally construed in *Chester,* Lord Hoffmann suggested that he would be prepared to take an approach similar to that taken by the majority of the House in *Rees* in a non-disclosure case which, as argued earlier, could be interpreted as recognizing the intrinsic value of autonomy. His Lordship was motivated to find against Miss Chester because on the ordinary principles of tort law the defendant was not liable. Nevertheless, he questioned whether a special rule should be created whereby doctors who fail to warn of risks should be made insurers against those risks.[129] According to Lord Hoffmann, such a rule would vindicate the patient's right to choose for herself and would recognize that the failure to disclose was an 'affront to her personality and leaves her feeling aggrieved'.[130] His Lordship could see that there might be a case for a modest solatium in such cases.[131] Although his Lordship was suggesting that compensation might be awarded for the 'affront to the claimant's personality', rather than actual physical damage, it seems that he still considered the materialization of the physical risk to be a crucial element of the claimant's right to a remedy. Insurers do not pay for risks that do not materialize. Nevertheless, it might be argued that the

---

129   *Chester v Afshar* [2005] 1 AC 134, Lord Hoffmann, 147.
130   Ibid.
131   Ibid.

recognition that compensation should be awarded for the 'affront to personality' as opposed to the physical injury is a step towards recognizing that the real harm is the failure to respect the claimant's ability to make use of relevant information in forming desires and making choices, as opposed to the unfortunate, but blameless, medical mishap.[132]

In terms of protecting autonomy, the actual result in *Chester* is strange. It could be argued that Miss Chester did not lose anything of value in terms of obviating the risk because she admitted that she would have run that risk in any event. However, she did lose something of value in terms of not being treated as able to evaluate and make a choice based upon the relevant material. Jackson suggests that the real harm in these cases is the failure to give sufficient information to patients prior to medical treatment, and that a remedy based on the unfortunate materialization of the medical risk fails to recognize this.[133] She argues that the failure to recognize being inadequately informed as actionable damage leaves a significant gap in the law's protection of patient autonomy. Jackson maintains that an individual's right to make an informed choice may have been compromised, regardless of whether she also happens to suffer injury.[134] Thus, in Jackson's view, the requirement that the claimant prove a causal link between inadequate disclosure and further actionable damage fatally undermines the capacity of the tort of negligence to accommodate the real harm suffered by patients who have been given insufficient information prior to medical treatment.[135]

The problem in *Chester* is that the interference with the intrinsic value of autonomy did not lead directly to the interference with the instrumental value because Miss Chester would not have chosen to obviate the risk and, therefore, the risk was one that she would willingly have chosen to bear, rather than one which she only chose to bear as a result of the interference with her autonomy. This seems particularly strange when compared to *Rees*. Here the interference with the intrinsic value of autonomy led directly to the interference with the instrumental value. Nevertheless, Ms Rees recovered a small sum which arguably reflected only the intrinsic value of autonomy. Of course, it was not quite as straightforward as this because the level of award did not appear to reflect the Lords' explicit views about where the value in autonomy lies, as evidenced by the fact that they clearly thought the consequences of the interference to be relevant even if they did not base the award of compensation on them.[136] However, in *Chester*, where the

---

132 P. Cane, 'A Warning about Causation' 115 (1999), *Law Quarterly Review* 21, 23.

133 E. Jackson, 'Informed Consent to Medical Treatment and the Impotence of Tort' in *First Do No Harm*, S. McLean (ed.), London: Ashgate (2006) 273, 274.

134 Ibid. 274.

135 Ibid. 282.

136 See earlier discussion of this point in *Rees*.

interference with intrinsic autonomy did not lead directly to an interference with the instrumental autonomy, the claimant received a significant sum effectively for the interference with instrumental autonomy.

It might be argued that the more pro-claimant result in *Chester* had something to do with the nature of the interference with autonomy. Mr Afshar's flippant response that he had not crippled anybody yet[137] to Miss Chester's question about paralysis could be interpreted as a deliberate,[138] rather than a negligent, interference with autonomy. However, their Lordships did not place emphasis on the fact that Miss Chester had asked a specific question regarding that risk.[139] Given this, we can assume that a claimant who is the victim of a careless, as opposed to an intentional, failure to disclose relevant information would be able to recover for the interference with the instrumental value of autonomy in the same way as Miss Chester. Thus, the more claimant-favourable result in *Chester* cannot be pinned on the notion of deliberate, as opposed to careless, conduct by the defendant.

Alternatively, it might be argued that the claimant-friendly result in *Chester* rests on the type of consequences which the court perceived[140] to have been a result of the interference with autonomy. The physical consequences which occurred in Chester, and which may occur as recognized risks of many operative interventions, are much more in line with the types of consequences for which the tort of negligence traditionally provides redress.[141] However, the damages for the birth of the child in *Rees* did not fit the traditional harm paradigm. Furthermore, given the focus on the paralysis in *Chester,* it is unlikely that the House of Lords would provide any protection for a failure to disclose a relevant risk that only interfered with the intrinsic value of autonomy because that risk does not eventuate. Thus, rather than providing comprehensive recognition of the interest in personal autonomy, this discussion demonstrates that in *Rees* and *Chester* the principle of autonomy functioned as a theoretical concept upon which their Lordships could hook the consequences which the claimant suffered so that they could, to varying degrees, provide damages in the tort of negligence. Proper protection of autonomy requires an analysis of what it is about autonomy that is so valuable.[142] If autonomy is of such significant value as the House of Lords suggested in *Rees* and *Chester*, then it follows that legal protection for autonomy should be comprehensive and consistent. Comprehensive protection would require legal recognition of

---

137   *Chester v Afshar* [2005] 1 AC 134, Lord Hope, 150.

138   Or reckless.

139   *Chester v Afshar* [2005] 1 AC 134.

140   I say this because, in theory, as I have argued above there was no interference with instrumental autonomy where Miss Chester would have willingly accepted the risk.

141   See above discussion.

142   N. Priaulx, *The Harm Paradox: Tort Law and the Unwanted Child in an Era of Choice,* Oxford: Routledge-Cavendish (2007), 184.

the whole value of autonomy. If the courts regard this as a step too far, they should at least be prepared to recognize the full extent of the consequences of the failure to respect a person's choice, irrespective of whether those consequences accord with the traditional physical harm paradigm. If the interest protected is autonomy, any consequences which arise as a result of the interference with that value should be compensable, whether they are physical or otherwise, because autonomy is not a value limited to interferences which have physical consequences.

Breach of autonomy and the hypothetical action

In the hypothetical saviour sibling scenario, there is a close link between the intrinsic value of autonomy and the instrumental value. This is because it is unlikely that the claimant will have a further opportunity to discover or obviate the consequences of the interference with autonomy before those consequences occur. In this respect, the hypothetical claim differs to the wrongful conception claims where the woman, arguably, has a further opportunity to choose whether or not to bear a child, because of the availability of abortion,[143] upon realization that she is pregnant.[144] It also differs from the non-disclosure cases where the instrumental interference with autonomy rests on whether a risk eventuates; that is, the eventuation of the risk is not an automatic consequence of the intrinsic breach. Nevertheless, this does not mean that both the intrinsic and the instrumental value of the parents' autonomy in the hypothetical claim cannot be recognized jointly or severally. The law could reflect the detriment to the parents that might occur from the knowledge that the reproductive project, that was so important to them, was treated with careless disregard by the medical team, in whom they had put their trust and who had undertaken to assist them in that project. It could also reflect the significant consequences that the failure to respect their desires will have on the advancement of their life plans and personal contentment.

An award designed to reflect the harm which occurs when one's autonomous reproductive choice is carelessly treated with disregard would be by way of a set amount. Given that English law does not recognize degrees of negligence, this sum would not vary between cases.[145] This sum would be

---

143 This is not to say that it is completely impossible for the parents to discover the negligent mistake before the birth of the child and then choose to have an abortion.

144 See, N. Priaulx, *The Harm Paradox: Tort Law and the Unwanted Child in an Era of Choice,* Oxford: Routledge-Cavendish (2007) for a comprehensive analysis of the concept of choice within the context of autonomy where questions of choosing abortion arise in wrongful conception.

145 The difficulty of ascertaining the amount of this sum might lead some to argue its inappropriateness. But, in theory, the amount of damages with respect to any physical injury which occurs through negligence is rather arbitrarily set.

available in all cases where the medical team carelessly transfers an embryo which does not meet with the parents' stated reproductive aim. Although, as noted above, the woman in the hypothetical claim is unlikely to have a further opportunity to discover the medical team's mistake and therefore obviate the consequences,[146] it is, in theory, possible for there to be an interference with the intrinsic value but not the instrumental value of autonomy in the hypothetical case. It is well known that pregnancy rates following IVF treatment are not particularly high.[147] Assume that the medical team discovers the careless mistake shortly after the 'wrong' embryo has been transferred to the woman. Upon identification of this mistake they inform the parents who are distraught and feel aggrieved that the medical team acted carelessly with respect to the project which was so important to them. Despite this, it is discovered at the pregnancy test two weeks after the embryo transfer, that the woman has not become pregnant. In this scenario no further consequences arise as a direct result of the negligent mistake.[148] Nevertheless, the fortuitous fact that a pregnancy does not occur may not alter the fact that the parents feel aggrieved, purely through the experience of having the entrusted medical team behave so carelessly with probably[149] one of the most important decisions of their life.

Instrumentally speaking, a host of consequences could, in theory, occur as a result of the negligent failure to respect the hypothetical parents' reproductive choice. A scenario similar to that just described above might occur, but where the woman does become pregnant. Here the woman might choose to have an abortion,[150] or she might choose to give the child up for adoption. It might be argued that this further choice opportunity should be seized by the parents to mitigate the effects of the original breach of autonomy. If the parents do not mitigate their loss, it might be argued that the defendant should not be responsible for the full consequences of her breach of autonomy. In response to this contention, Priaulx argues that presenting abortion

---

146   Having undergone PGD, there will be no clinical indication for the parents to undergo invasive and risky postnatal diagnosis. Furthermore, the particular disorder (currently largely haematological conditions), and certainly the tissue type, will not be identified during routine pregnancy screening tests and scans. Thus, preventing the parents from having a further opportunity to identify and obviate the effects of the medical team's carelessness.

147   Success rates may be lower in PGD because the number of embryos to choose from is diminished by the process of selection.

148   In theory, it could perhaps be argued that the woman now needs to undergo a further invasive procedure to replace one of the correct embryos and that she might have got pregnant in the first place with the correct embryo. However, there is of course no guarantee that this would have been the case. She could just as easily have failed to become pregnant with one of the correct embryos.

149   The plans of parents related to a cure for their seriously ill child, and birth of a subsequent child, will be of utmost importance in their lives.

150   N. Priaulx, *The Harm Paradox: Tort Law and the Unwanted Child in an Era of Choice*, Oxford: Routledge-Cavendish (2007). The issues which arise when considering autonomously choosing or not choosing abortion will not be considered here as they are comprehensively considered by Priaulx.

as a matter of choice ignores the fact that most women would rather not be in the position of having to make that choice at all.[151] This argument also applies in the context of adoption. Although the parents do not want another child who cannot save their existing sick child, perhaps particularly if the saviour child is also affected by the serious illness, they may feel a compulsion to keep the child rather than give the child up for adoption. That is, the parents' decision not to put the child up for adoption might be motivated by reasons other than an intrinsic desire to raise the child in and of itself. They might feel unable to assign the fate of the child to an unknown adoptive family, especially if they are lovingly raising other children. The parents' feeling that adoption is not a viable option may be intensified if the saviour child has the illness that also affects the child whom they were intending to save. The parents will have become adept in caring for a child with the particular illness and may doubt the ability of others who lack such experience to provide a similar standard of care. Furthermore, it might be argued that parent(s) who passed on the genetic condition to the child will feel some sense of responsibility for the child's illness and will therefore feel a sense of duty to provide care for the child themselves.

Nevertheless, for Priaulx, the fact that not choosing an abortion can be regarded as an autonomous choice does not exonerate the negligent medical practitioner from liability for the child's wrongful conception in wrongful conception cases. In these circumstances, Priaulx asks 'Is it inevitable in all situations that he who "chooses" must always take responsibility?'[152] By severing the link between responsibility and choice where the decision-maker has been forced into having to make the choice by another's wrongdoing, the wrongdoer retains responsibility for the consequences of the decision-maker's choice.

The force of the argument that the link between choice and responsibility ought to be severable is even stronger in the hypothetical case than it is in the wrongful conception scenario. In the hypothetical scenario the instrumental good, which the parents were seeking to achieve through the exercise of their autonomy, was a child who could save their existing child. Even though they might have a further opportunity to decide not to have or keep the child who cannot act as a saviour, this opportunity does not allow them to obtain that good which the original exercise of their autonomy sought to make possible. In the wrongful conception scenario the instrumental value of the original choice to be sterilized was that it would make it possible for the parents not to have (more) children. On the face of it, the woman has another opportunity to achieve this good through abortion.[153]

151 Ibid. 151.
152 Ibid. 158.
153 This, of course, is a simplistic account of the problem that ignores all the adverse consequences associated with the need to contemplate abortion that would not have arisen if the original autonomous decision had been treated with due respect.

However, in the hypothetical scenario the opportunity to choose whether to abort or give the child up for adoption cannot achieve the good of a saviour child, which was the good the parents' original autonomous decision sought to achieve. Thus, the effects of the original breach of autonomy cannot be obviated through the second opportunity to exercise autonomy. In this way, it might be argued that if parents should not shoulder responsibility for decisions where the possible instrumental effects are the same as those which they were prepared to decide upon in their original choice, then the parents certainly should not shoulder responsibility where the instrumental effects of the decision they are forced to take are not the same as those that they were prepared to decide upon in the original exercise of their autonomy. On this basis the careless medical team should not be able to transfer responsibility for the instrumental effects of their negligence if the parents choose to keep the child. However, as stated above, it is unlikely that the question of abortion would arise where the parents have availed themselves of PGD with the very aim of avoiding abortion.[154] Furthermore, the courts have regularly stated that the failure to pursue abortion or adoption should not be viewed as a *novus actus* in wrongful conception cases.[155] Thus, from a theoretical perspective, the consequences of the interference with the parents' reproductive choice would encompass the responsibility for raising and caring for the unwanted child.

In the event that the parents do keep the child, what are the instrumental effects of the team's negligence? This depends on whether the careless failure was to detect the genetic condition or the wrong tissue type.[156] If there was a failure to detect the wrong tissue, the new child would be healthy, but not able to act as a saviour to the older child. Here the parents have lost the valuable opportunity to cure the older child.[157] A legal system which recognizes the instrumental value of personal autonomy should provide an award which reflects the continuing responsibility of caring for that child. Moreover, the parents might now feel that they will have to resubmit themselves to the PGD process, which may have been unnecessary if their first attempt to procure a saviour sibling had not been carelessly thwarted. Thus, they might claim that the failure to obviate the need to undergo painful[158] PGD a further time is a result of

---

154   Following the PGD, they will probably have no reason to suspect that they are not carrying the healthy, saviour child they intended to carry.

155   See above discussion.

156   It could, in theory, be both.

157   Given that Professor Ajay Vora who treated Charlie Whittaker at Sheffield Children's Hospital puts the success of a stem cell transplant between tissue matched siblings at greater than 90 per cent success rate once the saviour child's cord blood is harvested, successful treatment is very likely.

158   Physically and emotionally.

the breach of their autonomous decision with respect to the original procedure.[159]

In the worst case scenario the careless disrespect for autonomy could lead to the birth of a second child with the same condition as the existing child. Here, the aims which have been frustrated are the cure for the older child and the prevention of the birth of a further child with the condition. If the instrumental value of autonomy is, as Young suggests, based on the promotion of well being and the prevention of harm,[160] it is difficult to see how a greater interference with instrumental autonomy could occur. If the courts are prepared to recognize that the value of autonomy is derived from what it makes possible, then the parents' award should reflect a continuing responsibility for caring for the older child who cannot now be cured, and for the sick created child who would not have been born. Furthermore, if an element of the benefit of PGD, which the parents could expect to achieve, was the need to obviate further PGD, it might be argued that the parents now need to procure two saviour siblings in quite a hurry.[161] This could involve two further cycles of IVF and PGD to procure two further pregnancies, or one risky twin pregnancy.[162] In this scenario the parents would have continuing responsibility for two children: one healthy and one with the condition in question whom they would not have had if their original decision to have a saviour sibling had not been negligently frustrated.[163]

Conclusion

In the face of the increasing demand for PGD to produce saviour siblings, it might be argued that it is only a matter of time until an attempt to create a saviour is carelessly frustrated. Currently, English law on wrongful conception

159 Within the consideration of the possibility of redress in the tort of negligence, the focus here would be on the pain and emotional suffering of undergoing another cycle of IVF and PGD because it is assumed that the NHS is funding the treatment. Currently, this is not routinely the case but there is evidence that the NHS is beginning to fund such treatment. See *Bionews*, 'NHS to Fund Saviour Sibling Treatment' (22 March 2004). Available HTTP: www.bionews.org.uk/page_11894.asp (accessed 16 June 2010). Currently, most PGD is likely to be privately funded with a cycle costing up to £25,000 at some clinics. In these circumstances where treatment is carelessly frustrated, it could have significant financial implications for the parents. I am grateful to Professor Emily Jackson for making this point. Where treatment is privately funded the parents may have an action in contract as well as tort, but I will leave consideration of this issue to a contract lawyer.

160 R. Young, 'The Value of Autonomy' 32 (1982), *The Philosophical Quarterly* 35, 36.

161 The cure for the sick child is usually needed fairly urgently in these cases and any significant delay caused by negligence could be fatal. Furthermore, the mother's advancing age could have an effect on her fertility, making the likelihood of successful treatment more unlikely. I would like to thank Professor Emily Jackson again for pointing out these important issues.

162 If any practitioner would be willing to assist the parents to conceive twin saviour siblings.

163 This might appear to be a farfetched scenario, but it might be argued that the will of parents to do anything possible to save the life and health of their children should not be underestimated.

and wrongful birth would recognize any extra-ordinary costs which might be associated with raising the saviour child if she has the condition which affects the existing child. However, this represents a very limited view of the loss that might occur when a parents' project to create a saviour sibling is negligently frustrated.

Building on existing academic argument that the real loss in wrongful birth and wrongful conception cases rests on the interference with the parents' autonomy, this discussion focuses on what is valuable about autonomy in the context of reproductive choice. It is argued that a true reflection of the loss suffered by the parents requires recognition of both the intrinsic, and the instrumental, value in reproductive autonomy. Unless, and until, the whole value of autonomy is recognized in this way, judicial claims that the English law of negligence protects against interferences with autonomy are premature.

Chapter 6

Genetic information
Failure to disclose a genetic risk

Introduction

Genetic information enables people to know about the future of their health with a modicum of certainty. Information regarding personal genetic risks may be extremely useful in making significant life decisions. Thus, where relevant genetic information about a person exists, but it is not disclosed to her, she may feel harmed by this failure. Currently, if she brought an action in negligence, English law would not recognize her harm.

Part I of this chapter considers how harm in the tort of negligence focuses on adverse physical sequelae. On this basis, the possibility of a novel lost chance argument in the event of a failure to disclose a genetic risk is considered. Given that the House of Lords has not been particularly receptive to the proposition that damages can be reformulated as lost chances in cases of physical harm, Part II presents the harm in the novel genomic claim as an interference with autonomy. This part then considers what is valuable about the protection of autonomy by way of the disclosure of relevant genetic information.[1]

Part III acknowledges that any recognition of harm and imposition of a corresponding duty to disclose on health professionals could interfere with patient confidence. From the perspective of the principle of rational autonomy, it is argued that having relevant genetic information about oneself is crucial to making informed decisions and, therefore, to one's autonomy. The discussion here focuses on how the desire of the would-be recipient to have the information might be rational whilst the desire of the tested person not to disclose might not be rational.

The final Part acknowledges that even if the law recognizes that a non-disclosure which interferes with autonomy is harmful, the health professional's failure to disclose is not currently a legal wrong. Here the focus is on

---

1 This part of the discussion draws largely on the previous discussions of the intrinsic and instrumental value of autonomy in Chapters 3 and 5.

previous arguments which the English courts have relied on to reject similar claims regarding the absence of a special relationship and policy concerns regarding defensive practice.

The problem: failure to disclose a genetic risk

Diverse physical[2] and psychiatric[3] conditions are being shown to have a genetic basis and genetics are also implicated in lifestyle choices which are not pathological.[4] Genetic research has paved the way for medicine to predict and, in some circumstances prevent, the condition with which the gene in question is associated. However, whilst genetic testing is becoming relatively easy, it remains difficult to prevent the fate that those genes predispose. Manifestation of genetic disorders might depend on a number of factors, including the interaction between genes and the environment.[5] Research into the manifestation and treatment of genetic disorders is still in the experimental stages.[6] Nevertheless, in some cases, where a person discovers that she is at risk of a genetic disorder there are well recognized ways in which she can reduce or eliminate that risk.[7] Furthermore, even where the risk cannot be eliminated, there may be value in being able to do all one can to reduce the risk even if the impact of such actions is thought to be minimal.[8] Alternatively, even where there is thought to be nothing that can

2   For example, genes are implicated in Alzheimer's: R. Tanzi and L. Bertram, 'New Frontiers in Alzheimer's Disease genetics' 32 (2001), *Neuron* 181; cancer L. M. Dong, J. D. Potter, E. White, C. M. Ulrich, L. R. Cardon, U. Peters, 'Genetic Susceptibility to Cancer: The Role of Polymorphisms in Candidate Genes' 299 (2008), *JAMA* 2423; obesity J. V. van Vliet-Ostaptchouk, M. H. Hofker, Y. T. van der Schouw, C. Wijmenga and N. C. Onland-Moret, 'Genetic Variation in the Hypothalamic Pathways and Its Role on Obesity' 10 (2009), *Obesity Reviews* 593; to name but a few.

3   For a general discussion of the genetic component of psychiatric illness see, for example, P. Willner, J. Bergman, D. Sanger, 'Editorial Behavioural Genetics and its Relevance to Psychiatry'19 (2008), *Behavioural Pharmacology* 371.

4   See, for example, S. H. Rhee and I. D. Waldman, 'Genetic Analysis of Conduct Disorder and Antisocial Behavior' in *Handbook of Behavior Genetics*, New York: Springerlink (2009), 455.

5   Epigenetics adds a whole new layer to genes beyond DNA. It demonstrates that single nutrients, toxins, behaviours or environmental exposures of any sort can silence or activate a gene. Environmental epigenetics concentrates on the impact a person's environment can have on her genes. Duke University Medical Center (2005, October 27), '"Epigenetics" Means What We Eat, How We Live And Love, Alters How Our Genes Behave', *ScienceDaily*. Available HTTP: www.sciencedaily.com / releases/2005/10/051026090636.htm (accessed 19 March 2010).

6   See Chapter 1.

7   A particularly good example of this is a mastectomy and an oopherectomy for women who possess the BRACA 1 genetic mutation.

8   Just because a particular course of action does not have a high chance of success does not mean that people will not want to pursue it. Lene Koch argues that it is rational for people to pursue medical intervention which has a very low chance of success. L. Koch, 'IVF – An Irrational Choice?' 3 (1990), *Issues in Reproductive and Genetic Engineering: Journal of International Feminist Analysis* 235.

be done to eliminate a particular genetic risk, people might want to know their risk so that they can prepare for the manifestation of that genetic condition, or can live their lives in a way which they find optimum in the light of the potential manifestation of that condition.[9]

Assume that a health professional discovers that a patient possesses a genetic mutation which puts her at risk of a particular genetic disorder. Assume further that given the mode of inheritance of the disorder there is a significant chance that other members of the tested person's family also possess the deleterious genetic mutation. Despite this, neither the patient nor the health professional passes the information on to those at risk. Assume that upon the discovery that the tested person and the health professional knew of the risk to her and did not inform her, the would-be recipient feels aggrieved. For the purposes of this discussion, let us assume that the aggrieved relative brings a claim in negligence against the health professional alleging that she was negligent in failing to inform her of the genetic risk.[10] Damage is the gist of the action in negligence.[11] Thus, the primary question which will arise with respect to the novel claim outlined here is: what damage has the claimant suffered?[12]

Genetic risks and lost chances

Where the genetic condition subsequently manifests in the would-be recipient, she might try to argue that the harm that she has suffered as a result of the failure to disclose the risk is the genetic condition itself because, if she had known about the risk, she could have avoided it. However, there are currently few genetic conditions which can be avoided.[13] Where there is a greater than 50 per cent chance that the claimant could have avoided the particular genetic condition if she had known about the risk, she could argue that the manifestation of the genetic condition itself constitutes harm.[14]

---

9 Which might be quite different to how she would otherwise have chosen to live her life.

10 The claim could equally be brought against the patient. However, the discussion of a tort claim against a family member is beyond the scope of this discussion.

11 As discussed in detail in Chapter 2.

12 The issues which might arise with regard to genetic privacy are not considered here. Graeme Laurie has written an interesting book which considers genetic privacy in this context. See G. Laurie, *Genetic Privacy: A Challenge to Medico-Legal Norms*, Cambridge: Cambridge University Press (2002).

13 A person can significantly reduce the risk of the manifestation of a genetic disorder associated with a genetic condition in breast cancer, associated with the BRCA 1 gene. American Cancer Society, Available HTTP: www.cancer.org (accessed 16 April 2010).

14 The causal issue relating to the reformulation of the damage would be subsequent to the question of whether the health professional owes a duty of care and whether she has breached that duty, neither of which are currently established in English law. In line with the aim throughout this book, the focus here is on the types of damage. In this part the focus is on the reformulation of damage as a lost chance.

Because causation is assessed on the balance of probabilities, harm which was made probable by the negligence is treated as being caused by it, and that damage, provided it is otherwise a recognized type of damage in negligence,[15] forms the gist of the action. The all or nothing nature of the balance of probability test is particularly problematic for claimants whose argument is that the defendant negligently failed to improve a pre-existing condition because, even in the absence of the negligence, there was always a chance that the claimant would manifest the condition. Where there are other potential causes of the adverse outcome, the court assesses the risk created by the combined non-negligent causes and compares this with the risk created by the defendant's breach of duty. Problems arise for the claimant where, although the defendant made a significant contribution to the risk, it was less than 51 per cent.

For the most part, it is not currently possible for the claimant to argue that if the information regarding the genetic risk had been disclosed there was a greater than 50 per cent chance that she could have avoided the manifestation of the genetic disorder because treatment for genetic disease has not reached this stage. However, the claimant might try to reformulate the damage as the lost chance of avoiding the manifestation of the genetic disorder, as opposed to the genetic disorder itself. The problem is that the House of Lords has not reacted favourably to the reformulation of damage into the lost chance of avoiding personal injury.

In *Hotson v East Berkshire HA*,[16] the claimant sustained a hip injury after falling from a tree. The hospital negligently delayed the diagnosis of the injury and the patient suffered necrosis of the hip joint, which led to a permanent deformity. The innocent injury gave him a 75 per cent chance of suffering the deformity. However, the negligent delay in diagnosis carried a 25 per cent chance of causing the deformity. Rather than presenting the harm as the deformity itself, which he could not prove the hospital had caused on the balance of probabilities, he formulated the gist of his damage as the 25 per cent lost chance of recovery, which he could prove the hospital had caused on the balance of probabilities.

Thus, the question for the courts was whether loss of a chance could amount to actionable damage. The Court of Appeal argued that the categories of actionable damage should not be closed and concluded that pure loss of chance ought to be recognized as legally actionable damage.[17] Thus, once the claimant demonstrated that he had lost a chance, he was entitled to damages proportionate to that lost chance. However, the House of Lords took a traditional approach to damage and assessed whether the claimant

---

15 I.e. the harm of physical injury through the manifestation of genetic disease.
16 *Hotson v East Berkshire HA* [1987] AC 750.
17 Ibid. Donaldson MR, 296.

could demonstrate, on the balance of probabilities, that the defendant caused the adverse outcome itself. As there was a 75 per cent chance that the fall had already caused the necrosis and deformity, it was more probable than not that it had been caused by the fall. Thus, the Lords drew the inference that at the time of the negligent conduct, the necrosis and the deformity were already inevitable.

Recently, the House of Lords has confirmed that a lost chance of avoiding personal injury[18] cannot amount to damage in negligence. In *Gregg v Scott*,[19] a GP negligently failed to diagnose his patient's cancer with the result that treatment was delayed by nine months. The rare nature of the claimant's cancer made it difficult for the medical experts to say what, if any, adverse consequences were caused by the delay. Expert evidence estimated that if treatment had begun promptly[20] he would have had a 42 per cent chance of total recovery, but when treatment did begin he only had a 25 per cent chance of total recovery. The claimant argued that he should be compensated for the loss of expectation of life.

The majority of the House of Lords rejected the claim. Baroness Hale thought that it would be problematic if claimants were able to reformulate the gist of their claims as lost chances because 'almost any claim for loss of an outcome could be reformulated as a claim for loss of a chance of that outcome.'[21] The minority thought that Mr Gregg had lost something of value, which the law ought to recognize in the loss of the expectation of life.[22] Nevertheless, they distinguished the case from *Hotson* on the basis that past events should be treated differently to future probabilities.[23]

The House of Lord's fairly persistent refusal to recognize the reformulation of the damage as loss of a chance in personal injury cases might seem strange in the face of the fact that the lost chance approach has succeeded in relation to economic loss. A lost opportunity to litigate a claim,[24] a lost opportunity of gaining employment[25] and the lost opportunity to negotiate a better business deal[26] are recoverable. Thus, the problem does not appear to be with lost chance per se, rather something to do with extending this head

---

18 In the form of manifestation of disease.
19 *Gregg v Scott* [2005] 2 AC 176.
20 I.e. at the time of the negligent diagnosis.
21 *Gregg v Scott* [2005] 2 AC 176, Baroness Hale, 233.
22 Ibid. Lord Hope, 201; Lord Nicholls, 222.
23 Ibid. Lord Nicholls 185. Lord Hope thought Mr Gregg's case was really a case of proved physical injury (rather than loss of a chance) because the delay in diagnosis had led to the enlargement of the tumour, 202–205.
24 *Kitchen v Royal Air Force Association* [1958] 1 WLR 563.
25 *Spring v Guardian Assurance plc* [1995] 2 AC 296.
26 *Allied Maples Group v Simmons & Simmons* [1995] 1 WLR 1602.

of liability to personal injury.[27] This point has not escaped the Lords. Hotson argued that there was an analogy between his lost chance and the lost chances in the economic cases. Despite admitting that this analogy was superficially attractive, Lord Bridge concluded that there were formidable difficulties in the way of accepting it.[28] Similarly, recognizing the same distinction between these two types of case in *Gregg*, Baroness Hale referred to Tony Weir's explanation of the differences between lost economic chances and lost chances to avoid personal injury[29]:

> Where the claimant is suing in respect of personal injury or property damage, he must persuade the judge that that injury or damage was probably due to the defendant's tort, whereas in cases of financial harm it is enough to show that the claimant had a chance of gain which the defendant has probably caused him to lose. There is nothing irrational in this, unless one supposes that it is sensible to speak of 'loss of a chance' without saying what the chance is of. Losing a chance of gain is a loss like the loss of the gain itself, alike in quality just less in quantity: losing a chance of not losing a leg is not the same thing as losing a leg.[30]

Thus, currently it seems fairly settled that lost chances to avoid an adverse medical outcome do not amount to damage in the tort of negligence. Given this, the claimant in the hypothetical action is unlikely to be successful in arguing that her lost chance of avoiding the manifestation of a genetic risk should be recognized as damage in negligence.

An interest in autonomy as the basis for harm in the failure to disclose genetic risks

Given that the House of Lords has rejected the idea that lost chances of avoiding personal injury can amount to harm, let us focus on a different interpretation of the harm in the event of a failure to disclose information about personal genetic risks. Whatever the implications of not knowing this information, the individual might feel that the failure to inform her of the personal genetic risk is an interference with her autonomy, because knowing available relevant information is in itself important in being an autonomous person, forming autonomous desires and making autonomous decisions. If the harm in failing to disclose genetic information lies in the interference with autonomy, it is irrelevant whether the genetic disease has, or will,

---

27 W. V. H. Rogers, *Winfield and Jolowicz Tort*, seventeenth edition, London: Sweet and Maxwell (2006), 289.

28 *Hotson v East Berkshire HA* [1987] AC 750, Lord Bridge, 782.

29 *Gregg v Scott* [2005] 2 AC 176, Baroness Hale, 232.

30 T. Weir, *Tort Law*, Oxford: Oxford University Press (2002), 76.

manifest, or the extent to which it can be prevented. The harm lies in not having the relevant information about oneself.

The principle of autonomy forms the basis of the doctrine of informed consent. From a legal perspective, it has long been recognized that respect for a person's autonomy requires her to be informed of relevant risks before she consents to a medical intervention. Although there has been significant disagreement about whether the standard that the law requires with respect to disclosure of medical risks is one of the reasonable doctor or the reasonable patient, recently there has been a trend in favour of the latter, more claimant-orientated, standard.[31] Thus, having information which is relevant to medical decisions is an important aspect of autonomy, and the law recognizes the harm in being deprived of that relevant information.[32] Chapters 3 and 5 explore the value which is derived from autonomy. In particular, they consider whether the interference with autonomy in and of itself could be the basis for legally recognized harm, and whether harm to autonomy is based on further adverse consequences.[33] This distinction is also crucial to the question of when harm might arise in terms of interference with autonomy with regard to the question of the failure to disclose genetic risks.[34]

From an intrinsic perspective, it might be argued that a person is harmed simply by the failure to disclose relevant information about her

---

31 In *Pearce v United Bristol Healthcare NHS Trust* [1999] 1 PIQR 53, Lord Woolf, 59, said:

> if there is a significant risk which would affect the judgment of a reasonable patient, then in the normal course it is the responsibility of a doctor to inform the patient of that significant risk, if the information is needed so that the patient can determine for him or herself as to what course he or she should adopt.

Recently in the House of Lords in *Chester v Afshar* [2005] 1 AC 134, Lord Steyn, 143, appeared to adopt this standard. He said:

> how a surgeon's duty to warn a patient of a serious risk of injury fits into the tort of negligence was explained by Lord Woolf, with the agreement of Roch and Mummery LJJ in *Pearce v United Bristol Health Care NHS Trust*.

Thus, it seems that, with the express approval of Lords Hope and Walker, Lord Steyn changed the doctor's duty of disclosure from one based upon professional standards and *Bolam*, to one based upon the needs of the reasonable patient. Meyers notes that *Pearce* and *Chester* have dramatically changed the inquiry in medical disclosure cases from what physicians reasonably think the patient should be told to what reasonable patients in the circumstances would want to be told. D. Meyers, *Chester v Afshar*; Sayonara, Sub Silentio, Siddaway? In S. McLean (ed.), *First Do No Harm*, Aldershot: Ashgate (2006), 255, 264.

32 See *Chester v Afshar* [2005] 1 AC 134.

33 There is a consideration of *Chester* as compared with *Rees v Darlington Memorial Hospital NHS Trust* [2004] 1 AC 309 from this perspective in Chapter 5.

34 To avoid repetition, I do not revisit the theoretical argument regarding intrinsic and instrumental value of autonomy here. The discussion here builds on Chapters 3 and 5.

genetic risks. Once she discovers this failure, she might feel aggrieved on the basis that it amounts to a failure to treat her as a mature and capable adult who is able to cope with knowing information about her own genetic risks. Assuming that one's sense of self respect is related to the level of respect that she is afforded by others,[35] it might be argued that the failure to treat an individual as an autonomous person could interfere with her sense of self-respect and dignity even if it does not impinge directly on any decisions she makes. From this perspective, it was argued in Chapters 3 and 5 that interference with autonomy results in harm, irrespective of whether any further adverse consequences flow from the failure to respect autonomy. Although it was suggested in the discussions in these earlier chapters that this interpretation represents the true recognition of the harm that might be occasioned by interference with autonomy,[36] it was also argued that the value of autonomy is partly derived from what it makes possible. It was suggested in Chapter 5 that if English negligence law were to adopt a notion of interference with autonomy as harm, it would focus on the value of autonomy in terms of what it might have made possible rather than the value of experiencing autonomy per se. Given the potential of gene therapy, let us focus on how a failure to disclose relevant genetic information might have adverse effects on significant life decisions of the would-be recipient.

There are many ways in which a failure to disclose relevant genetic information might be harmful to a person because not knowing this information prevents her from acting upon it. In this way, the harm caused by the interference with autonomy is akin to that which English negligence law recognizes with respect to non-disclosure of medical risks. It is well established that where a legally relevant risk is not disclosed to a patient, resulting in her consenting to an operation that she would not have consented to, where that risk eventuates she is entitled to damages to the extent of the harm which arose by her taking the risk which she would not have taken if her autonomy had been respected.[37] In the hypothetical genetic risk scenario, the consequence of the failure to respect the would-be recipient's autonomy

35 As Dillon notes: 'It is part of everyday wisdom that respect and self-respect are deeply connected, it is difficult if not impossible both to respect others if we don't respect ourselves and to respect ourselves if others don't respect us'. R. Dillon, 'Respect', The Stanford Encyclopedia of Philosophy (2003), substantially revised (2009). Available HTTP: http://plato.stanford.edu/entries/respect/ (accessed 23 April 2010).
36 See again the argument to this effect in Chapters 3 and 5.
37 Although this was not the exact factual situation in Chester v Afshar [2005] 1 AC 134 because Miss Chester admitted that she would have submitted to the same intervention at some point, it was implicit in the decision that she would have recovered the same damages if she would never have exposed herself to the risk.

by disclosing relevant genetic risks might be that she runs risks of manifest-ing the genetic condition, which she would not have run if she had been aware of the risk.

However, one of the problems with arguing that the interference with autonomy has harmed the individual because it has prevented her from being able to avoid a genetic risk is that for the most part genetic conditions are not curable.[38] Where the risk could not have been avoided in any event, the interference with autonomy has not prevented the individual from running the risk.[39] However, presenting a situation as a risk or no risk position is an oversimplification of the aetiology of genetic conditions and the factors which play a part in their manifestation.

Some genetic disorders are monogenic and others multifactorial. Mono-genic diseases are inherited from a single gene and can be dominant or reces-sive.[40] Over 50 per cent of dominant genetic disorders are late onset,[41] so an individual may spend a significant part of her life unaware that she will manifest a genetic disorder. Monogenic diseases, such as cystic fibrosis and haemophilia, are relatively rare, affecting 1 per cent of the human popula-tion.[42] Testing for many of these single inherited gene disorders is available, but curative treatment for them is not. The focus of research into treatment for single gene disorders is gene therapy, which aims to identify a malfunc-tioning gene and supply the patient with functioning copies of that gene. As discussed in Chapter 1, gene therapy is still largely at the clinical research stage.[43]

Where an individual had a significant chance of avoiding the manifesta-tion of the risk, it might be argued that failing to protect her autonomy is harmful because it prevented her from fulfilling her desire to obviate the risk. Assume that the would-be recipient manifests the genetic condition in question, although it is currently unlikely, in the future it might be open to the would-be recipient to argue that there was a high likelihood that, with treatment, she could have avoided the disorder in the event of an early warning. In these circumstances, the interference with autonomy frustrated

---

38 See the more detailed consideration of treatment for genetic disorders in Chapter 1.

39 But it could, of course, have still interfered with her intrinsic autonomy as discussed above.

40 It is estimated that over 8,000 human diseases are caused by defects in single genes: J. Kaplan, 'Genomics and Medicine: Hopes and Challenges' (2002) 11, *Gene Therapy* 658.

41 G. Laurie, *Genetic Privacy: A Challenge to Medico-Legal Norms*, Cambridge: Cambridge University Press (2002), 95.

42 L. Liu, Y. Li and T. Tollefsbol, 'Gene-Environment Interactions and the Epigenetic Basis of Human Diseases', 10 (2008), *Current Issues in Molecular Biology* 25.

43 See discussion in Chapter 1 and The Gene Therapy Clinical Trials Worldwide website, provided by the *Journal of Gene Medicine*. Available HTTP: www.wiley.co.uk/genetherapy/clinical (accessed 26 April 2010).

the avoidance of the condition.[44] However, the loss is presented in terms of the interference with autonomy, that is the inability to be able to choose to avoid the risk if the individual so desired, as opposed to the genetic condition itself. Thus, the damages awarded would reflect the interference with autonomy, as opposed to the direct physical aspects of the genetic condition.

As the argument here is that the harm lies in the interference with autonomy, as opposed to the manifestation of the physical condition itself, the existence of the harm is not reliant on the manifestation being a probability.[45] As noted above, from an autonomy perspective, harm may occur in the case of any discovered non-disclosure. However, even if the value of autonomy is derived from what it makes possible, it does not follow that in the genetic risk scenario there must be a likelihood of being able to avoid the manifestation of the genetic condition. People can make rational decisions to undergo treatments or undertake avoidance measures even though they know that the chance they will make any difference to the manifestation of the risk is low.[46] It might be argued that there is something valuable in having the opportunity to try all you can to prevent a genetic disease in order to be able to accept its manifestation. If you were not able to try all that you could,[47] you might always wonder whether the condition would have manifested if you had done all you could to avoid it. In this way, value is derived from autonomy in terms of being fully informed because possessing the relevant information is important when forming desires and making decisions. However, from an intrinsic perspective, any failure to provide the individual with relevant information could interfere with her autonomy, irrespective of what that information might have made possible.[48]

The major common diseases that affect people in developed countries, such as cancer, diabetes and cardiovascular disease involve susceptibility genes and the interaction of these genes with the environment or other factors. Such conditions are often termed multifactorial and their progression can be influenced by a number of common non-genetic factors, including diet, exercise, alcohol and drugs, and exposure to toxic chemicals or radiation.[49]

---

44 This assumes that once the particular disease has reached an advanced stage it is less readily treatable, which is the case for many common conditions which have some genetic element, such as cancer and cardiovascular disease.

45 As it was in the loss of a chance argument.

46 Furthermore, a person's instrumental autonomy might be interfered with even if she cannot do anything at all to halt, minimize or delay the onset of the condition, if the information would be relevant to her decisions about her life at all.

47 Because you were deprived of the information about the possible existence of the genetic risk.

48 As noted above, the discussion in this chapter focuses on the instrumental value of autonomy.

49 G. Laurie, *Genetic Privacy: A Challenge to Medico-Legal Norms*, Cambridge: Cambridge University Press (2002).

The predictability of these conditions is low compared with monogenic conditions,[50] but research into treatments for multifactorial genetic conditions currently far outweighs research into single gene disorders.[51] Understanding of gene-environment interactions enables advice to be targeted at susceptible groups who might be able to adapt their environment to minimize the chance of manifesting a genetic disorder.[52] So, although it might not be possible to cure a genetic condition by replacing the faulty gene with a healthy one, it might be possible to minimize the risk that the genetic condition will manifest, or delay, its onset, or attenuate its severity with lifestyle changes. In these circumstances there might be only a small chance that the modification of one's lifestyle will impact on the manifestation of the condition. However, the interference with autonomy deprives the would-be recipient of being able to make an informed decision to avoid the environmental risk factors. The harm is in the failure to be able to make an informed decision, rather than in the manifestation of the condition itself.[53] As research into gene therapy and gene-environment interactions progresses, it is likely that more effective treatments and means of minimizing the onset of genetic conditions will be discovered. In the face of real possibilities to avoid the manifestation of a genetic risk, the ability to be able to make an informed decision whether to take steps to minimize the risk raises different possibilities. Whilst currently the protection of autonomy via the disclosure of relevant genetic risks might make it possible for a person to feel she has done all she can to prevent the manifestation of the disorder, in the future there might be a real possibility that if she knew about the risk she would have been able to avoid the disorder. The interference with autonomy remains the same, but the implications of the interference are different, and this might be taken into account with regard to the level of damages awarded.

The consequences of an interference with autonomy caused by a non-disclosure where there is a possibility, however small, of preventing, minimizing or delaying the manifestation of the genetic condition are fairly clear in terms of the opportunity the would-be recipient has in choosing to avoid the genetic condition in question. But a failure to disclose relevant genetic information still interferes with the would-be

50 See Chapter 1.

51 The Gene Therapy Clinical Trials Worldwide website, provided by the *Journal of Gene Medicine*. Available HTTP: www.wiley.co.uk/genetherapy/clinical (accessed 26 April 2010). The branch of genetic science which studies genetic traits related to the response to environmental substances is sometimes termed ecogenetics. See G. E. Pence, *Classic Cases in Medical Ethics*, second edition, New York: McGraw-Hill (1995), 407–408.

52 Such environmental adaptations are likely to have an effect only before the condition has become manifest, making early warning particularly important.

53 Similarly to the situation above where gene therapy only has a very small chance of success.

recipient's autonomy, even if there is nothing that she can possibly do to prevent, minimize or delay the onset of the genetic condition. Indeed, as discussed above, from an intrinsic perspective, the value of autonomy lies in the experience of being treated as an autonomous person who is capable of coping with relevant genetic information; it is not only relevant to questions of treatment. From an instrumental perspective, information about the trajectory of one's health might be relevant to many life decisions, such as whether and when to have children, what career to pursue, whether to spend one's savings or conserve them for old age or infirmity. Although the focus in this scenario is on the value of autonomy in terms of what it makes possible by the way of minimizing genetic disease, there is also value derived from autonomy where there is no cure because it makes it possible for the individual to tailor important life decisions to significant risks of great consequence.[54]

The conflicting interests of the tested person and the would-be recipient

One of the major drawbacks to the argument that the harm in this novel claim rests on the interferences with autonomy is that the protection of the would-be recipient's autonomy might conflict with the protection of the tested person's confidence.[55] Health professionals owe a well recognized duty of confidentiality to their patients.[56] However, it is also well recognized that this duty is not absolute. The law and professional ethical codes allow doctors to displace confidentiality in the public interest.[57]

---

54 The counter argument that many people would not want to know about such risks might be raised. This issue will be considered in detail from the autonomy perspective in the following chapter.

55 In Australia, guidelines issued under s. 95AA of the Privacy Act provide information for health professionals regarding the use and disclosure of genetic information to a patient's relatives where the patient does not consent to disclosure. The guidelines note the importance of the 'ethical obligation to maintain the confidentiality of information about patients and the aim to maintain respect, as far as is possible, for the autonomy and confidentiality of the patient and the genetic relatives'. However, they do not examine how these interests might be ordered and protected in the face of conflict.

56 The concept goes at least as far back as the Hippocratic Oath. Today the doctor's duty of confidentiality is defined by the GMC. Extensive new guidance on confidentiality was published in October 2009. Available HTTP: www.gmcuk.org/guidance/news_consultation/confidentiality_guidance.asp (accessed 10 May 2010). The Data Protection Act 1998 and the Human Rights Act 1998 also protect patient confidentiality. However, both Acts provide for legitimate disclosure in specified circumstances. These Acts are not considered in the discussion in this chapter, which focuses on the common law. For a book which does consider these issues, see G. Laurie, *Genetic Privacy: A Challenge to Medico-Legal Norms*, Cambridge: Cambridge University Press (2002).

57 *Attorney General v Guardian Newspapers (No 2)* [1988] 3 WLR 776, Lord Goff, 807.

Where disclosure might obviate physical harm the courts have sanctioned disclosure.[58] The General Medical Council (GMC) also recognizes that the duty to maintain confidentiality might be overridden by the need to protect public safety where the information in question concerns a patient's HIV status. The GMC publication, *HIV infection and AIDS: The Ethical Considerations*, provides that: 'A doctor may consider it a duty to seek to ensure that any sexual partner is kept informed.'[59]

Although it appears to be well recognized that the doctor's duty of confidence to her patient is trumped by countervailing policy considerations concerning the physical safety of others, the interest in autonomy is unlikely to trump the interest in having confidence maintained. However, in the non-medical context, the courts have been willing to displace confidentiality with respect to non-physical threats. It seems that in relation to the publication of personal information in the press, there has been a significant relaxation of the public interest defence.[60] In *A v B plc and Another*, Lord Woolf said:

> In many of these situations it would be overstating the position to say that there is a public interest in the information being published. It would be more accurate to say that the public have an understandable and so a legitimate interest in being told the information.[61]

The Court was willing to take a more relaxed approach to the question of whether there were grounds for disclosure in the public interest, which did

---

58 See, in particular, *W v Egdell* [1989] 1 All ER 1089; *R v Crozier* (1990) 8 BMLR 128 where the risks of harm which might be caused by detained individuals, if they were released, were speculative. See also *Re C (A Minor) (Medical treatment)* (1991) 7 BMLR 138, which concerned information about a mother's fitness to be a parent. There are also long standing statutory exceptions to the duty to maintain confidentiality in relation to some contagious conditions. See, the Public Health (Control of Disease) Act 1984.

59 General Medical Council, *HIV Infection and AIDS: The Ethical Considerations* (1988).

60 This book does not consider the prospect of an action in tort for breach of privacy. Other commentators have considered the disclosure of genetic information for the privacy perspective. See, in particular, G. Laurie, *Genetic Privacy: A Challenge to Medico-Legal Norms*, Cambridge: Cambridge University Press (2002). Furthermore, the argument that a tort of invasion of privacy should be created was unanimously rejected by the House of Lords in *Wainwright v Home Office* [2004] 2 AC 406. The closest the courts have come to a standalone common law tort of privacy can be found in the judgment of Sedley J in *Douglas v Hello* [2001] QB 967, 1001. However, in *Wainwright*, Lord Hoffmann interpreted Sedley's judgment to be no more than a plea for the extension of the action for breach of confidence and his dictum certainly did not support an abstract principle of privacy: *Wainwright v Home Office* [2004] 2 AC 406, 422. See also, R. Brownsword, 'An Interest in Human Dignity as the Basis for Genomic Torts' 42 (2003), *Washburn Law Journal* 413, 414; G. Phillipson, 'Transforming Breach of Confidence? Towards a Common Law Right of Privacy under the Human Rights Act' 66 (2003), *Modern Law Review* 726.

61 *A v B plc and Another* [2003] QB 195, Lord Woolf, 208. The case concerned footballer Gary Flitcroft's attempt to prevent information about his extra-marital affairs being published in a national newspaper.

not depend on there being a risk (speculative or otherwise) of physical injury. The public policy grounds for disclosure in breach of confidence are unlikely to be relaxed to this extent with regard to medical information. However, a disclosure in order to protect the would-be recipient's autonomy would have clear personal relevance in a way which the information in *A v B plc and Another* did not. Interestingly, the English appellate courts are willing to rely on the public interest exception to allow the disclosure of medical information where this ostensibly displaced the information subject's confidence.

In *R v Department of Health Ex p. Source Informatics*,[62] the Court of Appeal held that there is no breach of confidence where the patient information has been anonymized, even if the reason for the disclosure is financial gain. A data base company wanted to sell information on GPs' prescribing habits from patients' prescriptions to drug companies, but Department of Health policy warned that this could amount to a breach of confidentiality, even if patient information was anonymized, because the information had been disclosed in confidence by the patients.[63] The Court unanimously held in favour of Source Informatics. Simon Brown LJ gave the leading judgment and held that because the duty to maintain confidence is based on equity, the question rested on whether the use of the information could be considered to be fair. According to Simon Brown, the use of the information by Source Informatics would not be unfair because the touchstone of unfairness was the conscience of the confidant.[64] Pattinson argues that:

> The conscience test, by modifying the scope of the duty of confidence, appears to make disclosure easier to justify than it would under the public interest test, which operates only where a duty of confidence exists. The upshot is that the conscience test is more likely to lead to patient's interests being overridden.[65]

Pattinson predicts that the conscience test could also be relied on to justify the disclosure of identifying information. He continues:

> the Court also stated obiter that anonymisation was not always required for use within the NHS for research and management purposes.[66] Many such uses of patient information were said to be potentially justifiable

---

62   *R v Department of Health Ex p Source Informatics* [2001] QB 424.

63   Ibid. Simon Brown LJ, 431.

64   Ibid. Simon Brown LJ, 439, felt that the reasonable pharmacist's conscience would not be troubled by the proposed use of the patient's prescriptions.

65   S. Pattinson, *Medical law and Ethics*, second edition, London: Sweet and Maxwell (2009), 225.

66   *R v Department of Health Ex p. Source Informatics* [2001] QB 424, Simon Brown LJ, 444.

in the public interest or because the scope of the duty of confidentiality does not extend to them. This latter approach would, presumably, involve relying on the conscience test, whereby one asks whether the conscience of a reasonable person *in the position of the confider* would be troubled by the use of identifying data for the purpose in question.[67]

Thus, it seems that the duty to maintain confidence is not absolute and, at times, the court might be willing to construe the ambit of the public interest or conscience defence widely if they see merit in the disclosure of the information.[68] Assuming that there is a well organized system for disclosing genetic information to relevant people which does not involve the tested person,[69] it is arguable that it might be possible for the tested person to remain anonymous. If the information could be effectively anonymized, it would no longer be subject to a duty of confidence. However, in many families this might be particularly difficult and would not provide a wholesale solution to the question of whether the would-be recipient's interest in knowing the information should supersede the tested individual's interest in the recipient not knowing it.

It might be argued that from the perspective of the interest which the duty of confidence is designed to protect, a duty to maintain confidence could yield in the face of the duty to disclose personal genetic risk.[70] In essence, the duty to maintain confidences arguably boils down to a protection of autonomy. Disclosure of otherwise confidential information will not be a breach of confidence in circumstances where the subject of the information consents to the disclosure.[71] Thus, it is not the information itself which is protected, but the subject's desire to keep that information from others. Furthermore, as Pattinson notes above, it might be that under the *Source Informatics* conscionability test, if the perception is that ordinary people would not be troubled by the disclosure of identifying information, then it may not be a breach of confidence to disclose that

---

67 S. Pattinson, *Medical law and Ethics,* second edition, London: Sweet and Maxwell (2009), 225 (original emphasis).

68 The aim here is not to provide a comprehensive discussion of the interference with confidence that might occur in the face of a duty to warn people about genetic risks. The aim is to demonstrate that the duty of confidentiality is not absolute and, therefore, that there is scope for it to be overridden by countervailing autonomy interests.

69 However, as we will see below, many people who are tested are themselves willing to speak to relatives who are at risk.

70 McLean argues that the tradition of confidentiality and privacy is potentially threatened by genetic information. S. McLean, *Autonomy, Consent and the Law* London: Routledge-Cavendish (2010), 180–182.

71 See, for example, *C v C* [1946] All ER 562. There is also no breach of Article 8 of the Human Rights Act 1998 if the claimant consents to disclosure of information about herself. See, for example, *Z v Finland* (1998) 25 EHRR 371, paragraph 112.

information. Where the subject of the information has consented to its disclosure, it is irrelevant how sensitive the particular information is objectively considered to be. Indeed, people might consent to the disclosure of all sorts of sensitive information about themselves because they do not object to others knowing it.[72] Thus, in the hypothetical scenario the issue can be framed in the following way: is the tested individual's interest in autonomy, i.e. her wish not to have her genetic information disclosed, outweighed by the would-be recipient's interest in autonomy, i.e. her interest in having relevant information about her genetic health disclosed to her?

From this perspective, the tested person's interest in autonomy is pitched against the would-be recipients.[73] Currently, the overriding interest, which incidentally protects the tested person's autonomy, is based on confidentiality and the fact that she has a pre-existing relationship with the health professional who would disclose the information. Thus, there is an existing duty of care, and in order to fulfil her professional confidentiality responsibilities the professional therefore protects the tested person's autonomy, even if she believes, intuitively or theoretically, that her autonomy should yield to the would-be recipient's. However, legal and professional duties regarding respect for autonomy could be determined on the basis of the respective value of the interest that the duty aims to protect, as opposed to the simple existence of a pre-existing relationship which might bear no relevance to the importance of the value protected.

72 People seem to be becoming less worried about others knowing their personal medical information. On television programmes, such as Embarrassing Bodies on Channel 4, people share their medical conditions which might typically be regarded as embarrassing and unsightly with, potentially, the entire nation. The show is particularly explicit, but it seems there is no shortage of people willing to sign up. The show's website is currently seeking people to make a new series, stating 'We're looking to help any patients who are *mortified about their problems* (my emphasis); we want to hear from you if you have an embarrassing physical problem that needs help'. They are particularly keen to hear from people who have 'been too embarrassed to seek help from [their] GP'. Available HTTP: www.channel4embarrassingillnesses.com/ (accessed 5 January 2010).

73 Where the interest in autonomy is protected in English law, the protection it receives is not absolute. It may yield to another's more important interest. That more important interest might be something other than autonomy, but it could also be a competing autonomy interest. For example, in *Pretty v UK* (2002) 35 EHRR 1, the European Court of Human Rights accepted that Pretty's right to autonomy, protected by Article 8, was interfered with, but that the interference with her autonomy could be justified under Article 8 (2). Like the House of Lords, the European Court felt that the restriction of Pretty's right to autonomy was necessary to protect the autonomy of vulnerable people. According to Lord Bingham in the House of Lords in *R (on the application of Pretty) v DPP* [2002] 2 AC 800, 824, it was not hard to imagine that an elderly person, in the absence of any pressure, might opt for a premature end to life if that were available; not from a desire to die or a willingness to stop living, but from a desire to stop being a burden to others. Thus, Pretty's interest in autonomy in being able to request assistance in suicide was overridden by vulnerable people's interest in autonomy in not being pressured into opting for suicide when this is not what they really want.

A value rationality approach to the disclosure of genetic risks

In order to consider the competing autonomy interests of the tested person and the would-be recipient, we can refer back to the detailed consideration of the principle of autonomy in Chapter 3 of this book. There, it was argued that rational autonomy, as opposed to a basic liberal interpretation of autonomy,[74] could form the basis of recognized harm in the tort of negligence. In the circumstances of this dilemma, the concept of rational autonomy might provide a theoretical basis for ordering the two competing autonomy interests. In Chapter 3 there was a discussion of different interpretations of rationality which might form an aspect of autonomy. A distinction was drawn between notions of rationality, which are premised on procedure, and those premised on particular substance or values. Within the context of the discussion of value rationality in Chapter 3, the difficulty in determining a legitimate definition of value in the context of conflicting values was discussed. Given the legal nature of the enquiry in this book, it was suggested that the court's established mechanism of pouring content into the notion of value, via the medium of the perceived values of the ordinary person, might provide one way in which the courts could determine what is substantively rational. However, in the context of this hypothetical dilemma, it might be argued that the values of the person who has put herself forward for genetic testing provide substance to the conception of rationality, which could form an aspect of autonomy.

As outlined above, the assumption is that a person approaches her doctor for a genetic test. This test shows her to be at risk of a genetic condition. The nature of this condition is such that many of the tested persons' relatives are at risk of manifesting the genetic condition. The tested person does not want her relatives to know this information because they will then know about her genetic risk and, therefore, she refuses to pass this information on to her relatives herself. We are not only assuming here that having the information would allow the would-be recipient significant potential to avoid or attenuate the manifestation of the genetic risk.[75] As discussed above, where the interest interfered with is presented in terms of autonomy, the failure to disclose any relevant information would prima facie interfere with a person's autonomy. It might be countered that people do not want to know information about their genetic risks and, therefore, giving them unsolicited information as to

---

74 Which is the interpretation of autonomy that the English common law explicitly adopts in medical law.

75 Although it might be the case that they can. See the BRCA 1 genetic mutation example given above.

these risks does not protect their autonomy.[76] This is a complex issue, to which the following chapter is dedicated. However, for the sake of argument, let us assume for the moment that the would-be recipient would very much welcome this information. Moreover, for the sake of this argument, the assumption here might be that autonomy is always protected by providing people with relevant information about their own health.[77] Indeed, on the face of it, it is not easy to appeal to the principle of autonomy to ground the interest in not having relevant information about oneself. If we assume for now that it is rational to want relevant information about genetic risks, one's interest in rational autonomy is protected when relevant information is disclosed. However, it is arguably not rational for a person not to want relevant genetic information to be passed on to relatives because this might identify her; thus, it is not a breach of her rational autonomy if this information is, in fact, disclosed. So why might a wish not to pass relevant genetic information on to relatives, because it might lead to the identification of one's own genetic traits, not be rational?

First, the rationality of the wish not to pass on this information might be questioned from the subjective perspective of the particular individual. She clearly believes it is important to have this information as evidenced by her decision to undergo genetic testing.[78] Given that the tested individual sees value in knowing this information about herself, it is arguably irrational for her not to want to pass this information on to others on the basis that it might allow them to identify her genetic status, especially since it is inherent in her knowing about her own genetic status that she knows about her relatives' genetic risks. Given that the tested individual wants to know this information, if she were the would-be recipient she would want to know the relevant genetic information about herself. Thus, on the basis of her own values, it is not rational for her to withhold the relevant information from others.

---

76 Or taking it one stage further, a sceptic might argue that it is not rational to want to know about medical risks that you can do nothing about. However, people might use medical knowledge about themselves to make all sorts of decisions which are not necessarily medical, such as reproductive, career or relationship decisions. In a related vein, Lene Koch undertook interviews with women who persevered with IVF treatment despite the very low and diminishing chance of successful treatment. She argues that even though their decision might be seen by many as irrational, given the risks and costs of IVF, for these women it was perfectly rational because it was not necessarily premised on the view that they would have a baby in the end, but on the fact that once they had tried everything they could move on to accept their infertility. L. Koch, 'IVF – An Irrational Choice?' 3 (1990), *Issues in Reproductive and Genetic Engineering: Journal of international Feminist Analysis* 235.

77 This issue is merely presented here, but not argued, because it forms the subject matter of Chapter 7.

78 Her decision to have testing might have been on the basis of a prior suspicion of a genetic risk as a result of a family history. However, it could also be an unknown risk which is picked up by one of the companies which is currently offering tests which analyse your DNA for many different traits and conditions in one go. Testing kits can currently be purchased online from deCODEme, 23andMe, International Biosciences and other companies.

In Chapter 3 the centrality of rationality in Kant's conception of autonomy was discussed. Guyer interprets Kant's conception of autonomy as: 'The property of the will by which it is a law to itself, or since law must be universal, the condition of an agent who is subject only to laws given by himself but still universal.'[79]

The achievement of autonomy is characterized by attaining a level of rationality where one can freely prescribe to oneself laws which are rational because they are valid for all rational beings. On this view, to be autonomous is to be rational in that one is motivated by reasons which are not particular to one's circumstances because to act otherwise would interfere with the categorical imperative to act only in accordance with that maxim, through which you can at the same time will that it become a universal law.[80] If the tested individual would want the relevant information to be disclosed to her, from the perspective of universalizability, it is not rational for her to prevent this information being passed on to relevant others. Given this, her desire not to disclose is not rationally autonomous.

Furthermore, evidence shows that, for the most part, when people discover themselves to be at risk of a particular genetic disorder and discover the implications of genetic information for other people, they willingly share the information with those for whom it has special significance.[81]

The Human Genetics Commission, which is the UK Government's advisory body on new developments in human genetics and their impact on individual lives, states: 'Experience shows that most people are willing to share (genetic) information with other family members once they are aware of the implications for those people.'[82]

Genetic Alliance UK, a support group for families with genetic disorders, also believes that it is better for individuals to be informed of their genetic risk than not to be so informed.[83] In taking this perspective, the Alliance enjoins families to be open about their genetic risks.

Furthermore, in a study of the views of women undergoing screening for the BRCA1 and 2 genetic mutations, the researchers noted that in justifying their participation in mutation searching, the women frequently cited the need to preserve others' autonomy, often at the expense of their own, as the very reason for undergoing genetic testing.[84] Thus, it seems that many

79 P. Guyer, 'Kant on the Theory and Practice of Autonomy' 20 (2003), *Social Philosophy and Policy* 70.

80 See K. Ansell-Pearson, 'Nietzsche on Autonomy and Morality: The Challenge to Political Theory' (1991) 39, *Political Studies* 270, 273; M. Boylan, *Basic Ethics,* New Jersey: Prentice Hall (2000), 94.

81 A. Sommerville and V. English, 'Genetic Privacy: Orthodoxy or Oxymoron' 25 (1999), *Journal of Medical Ethics* 144, 149.

82 Available HTTP: www.hgc.gov.uk/client/Content (accessed 16 April 2010).

83 Genetic Alliance UK Confidentiality Guidelines, London: Genetic Alliance UK (1998): 10.

84 N. Hallowell, C. Foster, R. Eeles, A. Ardern-Jones, V. Murday, M. Watson, 'Balancing Autonomy and Responsibility: The Ethics of Generating and Disclosing Genetic Information' 29 (2003), *Journal of Medical Ethics* 74.

people who discover themselves to be at risk of a genetic condition do want to disclose the information to relevant relatives.[85] If the position of ordinary people in the position of the tested individual is taken as the basis for rationality,[86] then it can further be argued that the decision of the tested individual to prevent relevant disclosure is not rational. The general attitude to sharing genetic information may mean that those who are not warned, because of their relative's reluctance, will be a small minority. The upshot of this might be that the would-be recipient feels particularly angry about the failure to disclose the relevant information to her, given that the accepted norm amongst affected parties is to disclose.[87]

Even if the interference with rational autonomy caused by a non-disclosure[88] is recognized by the law to be a harmful interference with the would-be recipient's autonomy, and her interest in rational autonomy outweighs any interest in autonomy that the tested person has, the imposition of liability with regard to this harm remains problematic because it does not result from a recognized legal wrong. Health professionals are not currently required to warn at risk relatives of genetic risks which have come to light because of a genetic test they have performed on a patient. Indeed, from the perspective of current legal values, it is possible that the health professional commits a wrong if she does disclose the information because, where the patient objects to the disclosure, it might amount to a breach of confidence, which is currently recognized as a legal wrong. Although English law recognizes that health professionals justify the displacement of patient confidence where they seek to protect countervailing public interest,[89] this is not the same thing as creating a duty to warn so that the health professional commits a wrong whenever she does not disclose the relevant information. The current indications are that if the health professional does not furnish a warning where there is a threat to the public interest, which occurs, she will not be held to have committed a wrong because there is a discretion to warn, but not a duty. Thus, if the hypothetical would-be recipient is to receive legal protection for the interference with her autonomy, it must follow that someone has a duty to disclose the information which, if not disclosed, would amount to an interference with her autonomy. As outlined above, this

85   For further support for this position see L. D'Agincourt-Canning, 'Experience of Genetic Risk: Disclosure and the Gendering of Responsibility' 15 (2001), *Bioethics* 231, 237–238 and I. Rabino, 'Genetic Testing and its Implications: Human Genetics Researchers Grapple with Ethical Issues' 28 (2003), *Science, Technology and Human Values* 365.

86   In Chapter 3 it is suggested that in the English legal context the content of the concept of rationality could be determined by reference to the court's perception of the ordinary person.

87   It might be that her anger is directed towards her relative, but it could also be directed towards the medical practitioner who was also aware of her risk.

88   Whether intrinsic or instrumental.

89   See, for example, *W v Egdell* [1990] 1 All ER 835; *R v Crozier* (1990) BMLR 128; and *Re C* (1991) 7 BMLR 138.

duty could be attributed to the tested individual or the health professional who possess the relevant knowledge. Here the focus is on the duty that might be owed by the health professional.

Is it wrong for medical practitioners not to disclose genetic risks? The implications of the imposition of a novel duty

There is likely to be some reluctance on behalf of the English courts to impose a duty on health professionals to disclose genetic information with the result that a failure to disclose would amount to a legal wrong.[90] In the absence of a special relationship, the English courts are currently unlikely to want to impose a duty of care with respect to a pure omission, especially if the duty might interfere with the health professional's other responsibilities, and this cannot be robustly justified by reference to the importance of the protection of the novel interest. This discussion focuses on the court's reliance on public policy grounds for refusing to recognize novel professional duties of care to warn third parties of impending risks; namely the risk that, in the absence of a special relationship, imposing such a duty will encourage defensive professional practice. For the sake of clarity, this discussion concentrates on the failure to disclose a genetic risk where the failure to disclose that risk interferes with the would-be recipient's autonomy because it forecloses options for treatment and avoidance.[91] If this approach to recognizing

---

90 As acknowledged above, the need to preserve confidence will lead to some resistance. This issue is not revisited here as it has also been considered by other commentators. See, for example, D. Wertz and J. Fletcher, 'Communicating Genetic Risks' (1987) 12, *Science, Technology, & Human Values* 60; H. Minkoff and J. Ecker, 'Genetic Testing and Breach of Patient Confidentiality: Law, Ethics, and Pragmatics' 198 (2008), *American Journal of Obstetrics and Gynecology* 498; M. Lacroix, G. Nycum, B. Godard and B. M. Knoppers, 'Should Physicians Warn Patients' Relatives of Genetic Risks?' 178 (2008), *Canadian Medical Association Journal* 593; R. Rhodes, 'Autonomy, Respect, and Genetic Information Policy: A Reply to Tuija Takala and Matti Häyry' 25 (2000), *Journal of Medicine and Philosophy* 114; M. Blake, 'Should Health Professionals be Under a Legal Duty to Disclose Familial Genetic Information?' 34 (2008), *Commonwealth Law Bulletin* 571; J. Kobrin, 'Confidentiality of Genetic Information' 30 (1983), *UCLA Law Review* 1283; G. Laurie, 'Obligations Arising from Genetic Information: Negligence and the Protection of Familial Interests' 11 (1999), *Child and Family Law Quarterly* 109; G. Laurie, 'Genetics and Patients' Rights: Where are the Limits?' 5 (2000), *Medical Law International* 25; G. Laurie, 'The Most Personal Information of All: An Appraisal of Genetic Privacy in the Shadow of the Human Genome Project' (1996) 10, *International Journal of Law, Policy and the Family* 74.

91 From the autonomy perspective then, the focus here is on instrumental as opposed to intrinsic value because this provides the best means of demonstrating why arguments regarding defensive practice should not prevent recognition of a duty. As the number of effective treatments grows, there is a greater potential for the imposition of a general duty to warn to decrease, rather than increase, the overall workloads of health professionals, on the basis of the assumption that prevention of disease is less labour intensive and less expensive than a cure. But this focus on instrumental value should not be taken as a reflection of the view that the intrinsic value of autonomy is not important, and should not similarly be protected.

a novel duty on the basis of interference of autonomy is accepted, it follows that any non-disclosure which interferes with autonomy should be capable of being subject to a duty of care, and the question of whether the value which rests on the consequences of the interference should be dealt with at the damages stage.

Tortfeasors are typically liable for making things worse, rather than not making them better.[92] It follows that English negligence law does not generally impose liability for pure omissions.[93] However, in some narrowly defined circumstances, the law does recognize that people owe a duty not to harm others by omission, particularly if there is some prior relationship between the parties.[94] Among such relationships is that between a health professional and her patient. In straightforward medical negligence claims the health professional owes a duty of affirmative action to those for whom she has undertaken to care. However, health professionals do not generally owe duties to prevent harm to those whose care they have not undertaken. The courts have been reluctant to impose a duty to warn on public service providers where it might interfere with the defendant's performance of her other responsibilities.

Public authorities are rarely held to owe a duty of care for damage caused by third parties because of a failure to control that third party, on the basis of the effect that the imposition of such a duty might have on the performance of the public authority's primary responsibilities. The most notable such claim is *Hill v Chief Constable of West Yorkshire*.[95] Here, the mother of the Yorkshire Ripper's last victim brought a claim against the police, alleging that they carelessly omitted to realize that Peter Sutcliffe, whom they had previously interviewed and released, was the killer. It was alleged that if they had conducted their investigations with due care and attention, Sutcliffe would have been apprehended before killing her daughter. The House of Lords held that the police force did not owe a duty of care. Lord Keith noted that the decision, in effect, made the police immune from this kind of action.[96] Nevertheless, his Lordship felt that such immunity was necessary because the imposition of a duty might encourage the police to perform their duties in a 'detrimentally defensive frame of mind',[97] thereby diverting police manpower and resources away from the police force's 'most important function – the suppression of crime'.[98]

---

92 J. A. Weir, 'Complex Liabilities' in K. Zweigert and U. Drobnig (eds), *International Encyclopaedia of Comparative Law*, Leiden; Brill (1976), 5.
93 *Smith v Littlewoods Organisation Ltd.* [1987] AC 241, Lord Goff, 271.
94 For example, the parent and child or employer and employee relationship. See also the discussion below with reference to special relationships regarding damage caused by third parties.
95 *Hill v Chief Constable of West Yorkshire* [1989] AC 53.
96 Ibid. Lord Keith, 64.
97 Ibid. Lord Keith, 63.
98 Ibid.

Similar concerns were evident in *Palmer v Tees HA*.[99] The mother of four-year-old Rosie Palmer, who was murdered by a psychiatric patient, previously cared for by the defendant health authority, sued the authority claiming that there had been a negligent failure to detect and treat the killer's violent tendencies, or otherwise prevent him from murdering her daughter. During his time as an in-patient at the hospital, the killer had stated that he had sexual feelings towards children and that a child would be murdered after his discharge. Mrs Palmer brought an action on Rosie's behalf, and one on her own behalf, alleging that the defendant's negligence had caused her psychiatric injury. Both claims were rejected. On striking out the mother's claim, the trial judge relied on the reasoning in *Hill,* and held that as a matter of public policy it would not be just, fair and reasonable to impose a duty of care in the circumstances because health professionals might be encouraged to engage in 'defensive practice in an effort to avoid liability, thereby neglecting their primary responsibilities of diagnosing and treating illness'.[100] The Court of Appeal upheld the decision.[101]

However, in one notable exception, the House of Lords did not rely on defensive practice concerns to protect a public authority from the imposition of a duty of care with respect to the actions of a third party. In *Home Office v Dorset Yacht Co. Ltd*[102] the House felt that the existence of sufficient proximity between three prison officers and the claimant yacht owner overrode defensive practice concerns. Thus, the House imposed a duty of care on the prison officers when they failed to prevent boys escaping from a borstal camp and damaging yachts. According to Lord Diplock, a special relationship, and therefore an element of proximity, existed between the yacht owner and the prison officers, by virtue of the fact that the prison officers had a special relationship with the boys and because the yacht owners were clearly exposed to a particular risk of damage should the boys escape. In the face of what they considered to be a proximate relationship, their Lordships concluded that there was no public policy reason, such as preventing defensive practice, for

99   *Palmer v Tees HA* [2000] PIQR P1.

100 Ibid. Gage J, 14. Interestingly, see the discussion in Chapter 2 regarding the court's reluctance to allow the defensive practice argument to prevent the imposition of a duty of care on public authorities in different circumstances. The discussion there focuses on the decisions in *Phelps v Hillingdon LBC* [2001] 2 AC 619 and *Barrett v Enfield LBC* [2001] 2 AC 550.

101 The Court of Appeal did not rely directly on the defensive practice argument. However, when the claim was put before the Court of Appeal, reliance on the third head of the *Caparo* test at the striking out stage was thought to be inconsistent with the decision of the European Court of Human Rights in *Osman v United Kingdom* (1998) 5 BHRC 293. However, it seems that it would now be open to the domestic courts to rely on this ground after the European Court subsequently admitted in *Z and Others v UK* [2001] ECHR 333 that its ruling in *Osman* was founded on a misunderstanding of English tort law and the procedural rules that allow the courts to strike out claims that disclose no reasonable cause of action.

102 *Home Office v Dorset Yacht Co. Ltd.* [1970] AC 1004.

failing to find the Home Office liable.[103] Thus, in the face of a special relationship, the defensive practice fear which was crucial to the failure of the actions in *Hill* and *Palmer*, where there was held to be no special relationship,[104] could not defeat liability.

In the hypothetical claim, the potential defendant cannot be said to exercise control over the patient. Thus, the defensive practice point might be more likely to be persuasive because where there is no duty to control, imposing one could divert manpower and resources towards a new duty which is not generally recognized as being part of the defendant's primary responsibilities.[105] But, new opportunities to prevent harm might call for the imposition of new duties which are not based on existing responsibilities of control, which import special relationships. Preventing new kinds of harm might entail the recognition of relationships which are not based on control, but are still sufficiently proximate not to be overridden by defensive practice concerns. In other words, what has traditionally been recognized as defensive practice might, in the context of changing social conditions,[106] come to be recognized as sensible cost-effective practice. On this basis, an element of control ought not to be essential to the finding of a special relationship between the health professional and the 'at risk' relatives which is required to defeat countervailing defensive practice concerns. Indeed, it makes no sense to insist on an element of control where, even if the defendant did have control over the actions of the patient, it would make no difference to the fact that the would-be recipient carries that genetic mutation because she acquired the fundamental susceptibility to that genetic condition at conception.[107] It might be argued that the basis for liability in these circumstances ought to rest on the health professional's ability to identify the party(ies) at risk.[108]

---

103   Ibid. Lord Reid, 1032.

104   In *Hill v Chief Constable of West Yorkshire* [1989] AC 53, Lord Keith, 62, felt that in the absence of any legal requirement to control the actions of the third party, there was no special relationship between the third party and the defendant which would allow them to conclude that there was a special relationship between the defendant and the claimant injured by the third party, which could override countervailing policy concerns. Similarly, in *Palmer v Tees HA* [2000] PIQR P1, Stuart-Smith LJ, 11, concluded that it was crucial to the rejection of the claimant's case that there was no special relationship between the defendant and the victim because, at the material time, the killer was an outpatient and therefore not in the 'custody of or under the control of the defendant'. Armstrong had previously been an in-patient at the defendant hospital. Pill LJ agreed with this point: *Palmer v Tees HA* [2000] PIQR P1, 16.

105   In *Home Office v Dorset Yacht Co. Ltd.* [1970] AC 1004, it was clearly part of the prison officers' responsibilities to prevent the boys from escaping.

106   Here, the increasing prevalence and importance of genetic information.

107   The issue of whether it would have made any difference to the progression of the relative's condition, even if the doctor had given a timely warning, will be dealt with in the following chapter.

108   Thanks are due to Derek Morgan for this point and for the many discussions he and I engaged in regarding the arguments in this book.

In the USA there is a fairly well developed body of case law addressing the failure to warn in cases of medical risk, in which an element of control between the defendant and the third party is not required to establish a duty of care. *Tarasoff v Regents of the University of California*[109] is the seminal US case which considers the health professional's duty to disclose confidential medical information to protect others. Prosenjit Poddar killed Tatiana Tarasoff. Whilst a voluntary outpatient at the University of California hospital at Berkley, Poddar had expressed his intention to kill a girl, readily identifiable as Tatiana, to his psychologist, Dr Moore. Upon Poddar's release, Dr Moore did not attempt to alert Tatiana or the local authorities of the impending peril. When Poddar killed Tatiana, her parents brought an action alleging inter alia that the therapists had a duty to warn Tatiana of the danger.

Tobriner J found that the common law imposed no general duty to warn those endangered by the conduct of a third party. However, in his view, an exception to this rule could be carved out where the defendant stood in some special relationship either to the third party or the foreseeable victim of the third party's conduct.[110] He concluded that although there was no special relationship between Tatiana and the defendant therapists, there was a special relationship between the therapists and Poddar, which could support an affirmative duty for Tatiana's benefit.

The English courts have required a special relationship between the claimant victim and the defendant before a duty of care can arise.[111] It has not been sufficient to simply demonstrate that there is a relationship of proximity between the defendant and the third party.[112] Moreover, even where there is a relationship between the claimant and the defendant, the courts have required that the defendant have some control over the third party before they will hold that there is a sufficiently proximate relationship to impose a duty on the defendant for the claimant's harm. The US courts have adopted a different approach, whereby the question of whether there is a special relationship importing a sufficient level of proximity does not depend on whether the defendant has any control over the third party per se, but on whether the defendant has any control over the protection of the victim. Thus, in essence, the imposition of liability rests on the victim being readily identifiable.[113] It is arguable that this element of identifiability provides the grounding of liability in these types of claim because it is identifiability, as opposed to control, that goes to the heart of an effective warning.[114] If the victim is readily identifiable and a warning will allow the victim to protect

---

109 *Tarasoff v Regents of the University of California* 551 P.2d 334 (Cal 1976).
110 Ibid. Tobriner J, 343.
111 *Home Office v Dorset Yacht Co. Ltd.* [1970] AC 1004.
112 See *Palmer v Tees HA* [2000] PIQR P1 and *Hill v Chief Constable of West Yorkshire* [1989] AC 53.
113 As Tatiana was.
114 That is, giving a warning to the victim might do nothing to control the actions of the third party.

herself, the defendant can influence the outcome of the course of events, even though she has no control over the third party.

A number of US courts have held that a duty to disclose confidential medical information might be imposed on doctors in order to prevent harm to third parties, even though the defendant had no physical or legal control over the third party. In *Davis v Rodman,* an Arkansas court held that where a doctor discovers that her patient has a contagious disease, she owes a duty to those who are ignorant of such disease, and, who by reason of family ties or otherwise, are likely to be in contact with the patient, to instruct them as to the character of the disease.[115] Taking things one stage further in *Bradshaw v Daniel,*[116] the Tennessee Supreme Court extended the doctor's duty to warn beyond cases where the patient is the source of the harm. The doctor treated a patient for Rocky Mountain spotted fever, which he had contracted when he was bitten by a tick on a camping trip. The court held that the doctor had a duty to warn the patient's wife that she was also at risk because she accompanied her husband on the camping trip. This duty was not dependant on any legal or physical right to control the actions of the patient as the disease did not emanate from the patient, but it was dependant on the identifiability of an at-risk individual.

Pate v Threlkel[117] was the first US case concerning the disclosure of a genetic risk. The Florida Supreme Court certified the following question as one of great public importance: 'Does a physician owe a duty of care to the children of a patient to warn the patient of the genetically transferable nature of the condition for which the physician is treating the patient?'[118] The Court found that the prevailing standard of care was developed for the benefit of the patient's children as well as the patient, and concluded that when an identified third party is at risk, and the physician knows of the existence of those third parties, then the physician's duty runs to those third parties.[119]

A year later, the Superior Court of New Jersey heard *Safer v Pack.*[120] Here, the claimant, who was the daughter of the defendant physician's patient, claimed that the physician had been negligent in failing to warn her about the genetic nature of her father's cancer. Kestin, JAD said:

> We see no impediment, legal or otherwise, to recognizing a physician's duty to warn those known to be at risk of avoidable harm from a genetically transferable condition. In terms of foreseeability especially there is no essential difference between the type of genetic threat at issue here

---

115 *Davis v Rodman* 227S.W. 612 (Ark. 1921).
116 *Bradshaw v Daniel* 854 S.W. 2d 865 (Tenn. 1993), 866.
117 *Pate v Threlkel* 661 So. 2d 278 (Fla. 1995).
118 Ibid. 279.
119 Ibid. 282.
120 *Safer v Pack* 677 A.2d 1188 (N. J. Super. Ct. App. Div. 1996). Also discussed above.

and the menace of infection, contagion or a threat of physical harm.... The individual or group at risk is easily identified, and substantial future harm may be averted or minimized by a timely and effective warning.... Although an overly broad and general application of the physicians duty to warn might lead to confusion, conflict or unfairness in many types of circumstances, we are confident that the duty to warn of avertible risk from genetic causes, by definition a matter of familial concern, is sufficiently narrow to serve the interests of justice.[121]

Thus, in the US, in the face of a readily identifiable at risk individual, an element of control over the patient is not required to establish that there is a sufficient relationship of proximity between the health professional who fails to warn, and the person who is aggrieved by the failure to warn.

Although providing a warning to third parties regarding potential genetic risks is not part of the health professional's primary responsibilities, if a duty to warn is imposed on health professionals it is arguable that they might spend more time ensuring that warnings are given to individuals at risk from genetic conditions at the expense of patient care. If the question of whether a duty of care might be imposed depends on a distribution between contributing to the maintenance of high standards and encouraging defensive practice which, in turn, depends on whether the action to which the duty relates is a primary responsibility, there is a significant chance that the courts will refuse to impose a duty in the hypothetical claim.

As noted at the beginning of this chapter, most medical conditions are now thought to have a genetic element. Thus, an undefined duty to disclose could result in health professionals spending a significant amount of time warning those at risk. However, this analysis fails to recognize the potential of modern genetics in promoting disease prevention. In the short term, the most widespread use of modern genetic technologies will be in diagnosis and screening, but the ultimate aim of research into human genetics is to prevent genetic disorders.[122]

For a number of reasons it is generally recognized in medicine that prevention is better than cure; in terms of successful treatment[123] and in terms of conserving stretched resources.[124] Moreover, where the preventative measure

121 Ibid. Kestin JAD, 1192.
122 House of Commons Science and Technology Committee, *Human Genetics: The Science and its Consequences*, Third Report (HMSO, 6 July 1995), 31–55, paras 65–124.
123 For example, a woman is less likely to die of breast cancer if she has a prophylactic double mastectomy than if she seeks treatment after the disease has become manifest.
124 See Department of Health, 'Securing Good Health for the Whole Population: Final Report', HMSO (25 February 2004). See also NHS Institute for Innovation and Improvement, *Prevention is Better than Cure*. Available HTTP: www.institute.nhs.uk/building_capability/technology_and_product_innovation/prevention_is_better_than_cure.html (accessed 23 June 2010).

involves simply abstaining from a particular behaviour or avoiding exposure to a particular toxin, the costs of that preventative measure will, from the National Health Service's perspective, be relatively low.

Health promotion is high on the NHS's agenda.[125] Promoting the nation's health involves educating people about healthy living and the avoidance of disease, and encouraging those groups who might be at risk from specific conditions to be screened so that conditions might be detected at the pre-disease stage.[126] Providing a warning to people who are known to be susceptible to a particular genetic disorder to enable them to avoid the behaviours that might influence the manifestation of their disease, is a step further along the path that emphasizes prevention over cure. It might be argued that one of the drawbacks to health promotion that has a general ambit is that people believe that the information is not relevant to them. If people are given health risk information based on personal genetic risk, they might be more inclined to see a personal link. Thus, a system which warns individuals of specific and personal genetic risks, where those risks might be avoided, might facilitate preventative medicine and promote the efficient use of NHS resources.[127]

This suggests that the imposition of a duty to disclose relevant genetic information to those at risk will not lead to the wasteful use of public resources. Furthermore, providing a warning to those at risk need not be as difficult from a practical point of view. The health professional need not personally seek out and warn the at-risk relatives. The duty could be fulfilled by warning the patient who could then convey the information to relatives. In the USA, where a duty of care has been imposed on doctors to inform third parties of genetic risks, the courts have reached different conclusions about what amounts to an effective warning.[128]

We know that for the most part patients want to inform their relatives of genetic risks.[129] Thus, the incidence of relatives failing to disclose relevant

---

125 One of the core aims of the NHS is to 'provide information services and support to individuals in relation to health promotion, disease prevention, self-care, rehabilitation and after care'. See NHS website: www.nhs.uk/England/AboutTheNhs/CorePrinciples.cmsx. See also J. McHale and M. Fox, *Health Care Law: Text and Materials*, second edition, London: Sweet and Maxwell (2007), 24–25.

126 Well men and well women clinics have been established to target gender specific risks.

127 The National Institute for Clinical Excellence (NICE) gives guidance regarding clinical medical treatment largely based on cost effectiveness: see M. Sculpher, M. Drummond, B. O'Brien, 'Effectiveness, Efficiency, and NICE' 322 (2001), *British Medical Journal* 943. NICE places significant emphasis on the importance of promoting good health and preventative medicine, indicating its recognition of the cost effectiveness of preventative measures, as opposed to secondary medicine: see National Institute for Clinical Excellence, *Social Value Judgments: Guidelines for the Institute and its Advisory Bodies* (January 2005).

128 In *Pate v Threlkel*, 661 So. 2d 278 (1995), the Supreme Court of Florida held that the doctor's duty to warn would be satisfied by notifying only the patient, 282. In *Safer v Pack* 677 A.2d 1188 (N.J. Super. Ct. App. Div. 1996), Kestin JAD, 1192, held that reasonable steps must be taken to ensure that the information reaches those likely to be affected or is made available for their benefit.

129 See the earlier discussion in this chapter.

genetic information is likely to be low, suggesting that if a duty to disclose were imposed on health professionals where relatives refused to disclose, it would not be unduly onerous.[130] Where the patient is willing to disclose the information to relatives, the health professional's direction to the patient regarding the need to disclose the information and verification of the patient's intention to do so could suffice to discharge the duty. Furthermore, evidence from a large, international survey of geneticists found that, where the manifestation of a genetic condition is avoidable, a majority of medical geneticists would disclose a patient's genetic information to a relative without the patient's consent.[131] Given that the imposition of a duty to disclose relevant genetic information is unlikely to impose an unduly onerous burden on health professionals, the courts should not rely on concerns about defensive practice to refuse to impose a duty of care where a duty would serve to protect the fundamental interest in autonomy.

Conclusion

The concept of being fully informed with respect to making important decisions is legally recognized with respect to medical care. The familial nature of genetic information is such that health professionals might possess information which is crucial to another's choices, actions and desires, whilst that other remains ignorant of that information. If having relevant information is an important part of one's autonomy, the failure to disclose that information interferes with autonomy. This is the case even where the would-be recipient of the information could not have relied on that information to try and avoid genetic disease. However, English negligence law does not currently recognize the harm that might be occasioned to the would-be recipient by way of interference with autonomy in these circumstances. Although the value of autonomy can be intrinsically or

---

130 In a twelve month prospective study investigating the frequency with which genetics professionals become concerned about the failure of clients to pass on such information to their relatives, a total of 65 cases of nondisclosure were reported, representing less than 1 per cent of the genetic clinic consultations in the collaborating centres during the study period. However, the study notes that practical barriers, such as geographical distance, family rifts, divorce, separation, adoption and large age gaps between siblings may impede communication. See A. Clarke *et al.*, 'Genetic Professionals' Reports of Nondisclosure of Genetic Risk Information within Families 13 (2005), *European Journal of Human Genetics* 556. However, retrospective surveys of members of the American Society of Human Genetics and/or American College of Medical Genetics and National Society of Genetic Counsellors found that a quarter of the clinical geneticists and half the counsellors reported having had patients who refused to inform family members at risk. See M. J. Falk *et al.*, 'Medical Geneticists' Duty to Warn At-Risk Relatives for Genetic Disease' 120 (2003), *American Journal of Medical Genetics* 374.

131 D. Wertz and J. Fletcher, 'An International Survey of Attitudes of Medical Geneticists Toward Mass Screening and Access to Results' 104 (1989), *Public Health Reporter* 35.

instrumentally construed, in order to appeal to legal sensibilities this chapter has focused on the harm that might occur, in terms of interference with autonomy, where the risk which was not disclosed would have been relevant to the would-be recipient's choices, actions and desires.[132] From the perspective of substantive rational autonomy, as referenced to the evident norms of those who have undergone genetic testing, it has been argued that the would-be recipient's interest in autonomy prevails over the autonomy interest of the tested individual.

Finally, it is argued that the courts should not fail to impose a duty of care in these circumstances on the basis that it would encourage defensive practice. On the contrary, the protection of the would-be recipient's autonomy in these circumstances has the capacity, in the longer term, to lessen the burden on the human and financial resources of the National Health Service.

---

132 I.e. the instrumental value perspective with particular emphasis on decisions regarding treatment and avoidance of genetic conditions.

Chapter 7

Genetic information

Unwanted disclosure of a genetic risk

Introduction

The previous chapter considered whether a failure to disclose relevant genetic information could amount to a legally recognized set back to the interest in autonomy for which a corresponding duty of care might be imposed on health care professionals. It was argued that individuals have an interest in autonomy and this interest would be protected by knowing relevant genetic information about themselves. However, the imposition of a wholesale duty to warn would fail to recognize that people can have an interest in not knowing relevant genetic information.

The fundamental problem with the interest in not knowing is that unless we ask people what they want to know, we are not aware of their desire not to know. Furthermore, any preliminary inquiry would make the existence and the essential character of the information known, so that the interest in not knowing is effectively interfered with. Knowing that one has an elevated risk of suffering a genetic disorder may well be distressing. However, upon current legal principles, knowing this information is unlikely to be recognized as legal harm and disclosing it is unlikely to be a legal wrong within the tort of negligence. On the whole, this book seeks to argue that the deleterious effects that might occur in each of the novel claims considered here might be recognized as harm in English negligence law if it were explicitly imbued with the recognition of an interest in personal autonomy. However, it is difficult to make a wholesale argument to that effect with respect to the interest in not knowing genetic information. If, as argued in the previous chapter, personal autonomy provides the basis for the interest in knowing information about oneself, it is difficult to also present it generally as the basis for the interest in not knowing the same information. Given this, in the context of the interest in not knowing genetic information, a distinction is made between unknown unknowns and known unknowns. Drawing on the discussion of autonomy in Chapter 3, Part I of this chapter demonstrates that, from a theoretical perspective, the interest in autonomy cannot serve as a basis for arguing that one has been harmed by the disclosure of relevant genetic information about oneself where the individual had made no express request not to know the information, i.e. prior to the disclosure it was an *unknown* unknown. Alternatively, Part II focuses on the situation where the relative actually indicated her desire not to be informed of genetic information but it was disclosed in any event, i.e. the disclosure concerned a *known* unknown. It is argued that in these circumstances there is significantly greater scope for arguing that disclosure interferes with the recipient's autonomy. This part then draws on the discussion in Chapter 3 to consider how Gerald Dworkin's theory of procedural autonomy might provide the basis for arguing that a failure to respect a person's express wish not to know genetic information might interfere with her autonomy.

Autonomy and the interest in not knowing an unknown unknown

From a liberal individualistic perspective of autonomy, which denotes largely unconditional self-determination, it is difficult to maintain the position that an interest in not knowing unknown unknowns, that is where the individual has not expressed a desire either to know or not to know the information because she has not considered the fact that there might be something to know, is grounded in autonomy because, the opportunity for self-determination had not arisen at the time of the disclosure.[1] Nevertheless,

---

1 There is a significant body of literature which addresses the interest in not knowing genetic information from the perspective of the breach of autonomy in the context of an explicit refusal to receive genetic information. Although many reject the argument that the interest in autonomy can provide the theoretical basis for the interest in not knowing at all, i.e. with regard to not knowing unknown unknowns and not knowing known unknowns. See, for example, J. Harris and K. Keywood, 'Ignorance, Information and Autonomy' 22 (2001), *Theoretical Medicine* 415; R. Rhodes, 'Genetic Links, Family Ties, and Social Bonds: Rights and Responsibilities in the Face of Genetic Knowledge' 23 (1998), *Journal of Medicine and Philosophy* 10; J. Wilson, 'To Know Or Not To Know? Genetic Ignorance, Autonomy and Paternalism' 19 (2005), *Bioethics* 492, others take a more nuanced approach arguing that although the interest in not knowing unknown unknowns cannot be justified by recourse to autonomy, the interest in not knowing known unknowns can. See, for example, K. Fulda, K Lykens, 'Ethical Issues in Predictive Genetic Testing: A Public Health Perspective' 32 (2006), *Journal of Medical Ethics* 143; G. Laurie, 'The Most Personal Information of All: An Appraisal of Genetic Privacy in the Shadow of the Human Genome Project' 10 (1006), *International Journal of Law, Policy and the Family* 74; G. Laurie, 'Obligations Arising from Genetic Information – Negligence and the Protection of Familial Interests' 11 (1999), *Child and Family Law Quarterly;* G. Laurie, 'Genetics and Patients' Rights: Where are the Limits?' 5 (2000), *Medical Law International* 25; G. Laurie, 'Protecting and Promoting Privacy in an Uncertain World: Further Defences of Ignorance and the Right Not to Know' 7 (2000), *European Journal of Health Law* 185; G. Laurie, *Genetic Privacy: A Challenge to Medico-Legal Norms,* Cambridge: Cambridge University Press (2002); J. Malek, L. Kopelman, 'The Well-being of Subjects and Other Parties in Genetic Research and Testing' 32 (2007), *Journal of Medicine and Philosophy* 311; R. Andorno, 'The Right Not to Know: An Autonomy Based Approach' 30 (2004), *Journal of Medical Ethics* 435. First, this discussion seeks to tackle the more vexed and lesser considered issue of the interest in not knowing unknown unknowns from the perspective of the interest in autonomy because in the current social context there is no clear system for registering one's wishes regarding the receipt of genetic information.

some commentators suggest that the interest in autonomy might 'imply an obligation not to give information to a person whether or not she asks not to be informed'.[2] Where it is not known whether the particular individual wants to know, the interest in not knowing could be said to be based on autonomy where the desire not to know can be based on some form of proxy decision-making which purports to simulate the decision the individual would have made.[3] However, in the absence of an explicit indication, it is not easy to know what others would want. Whilst human beings experience and perceive in a similar way to one another, there is something unique about experiences and perceptions.[4] Atkins argues that it is this uniqueness that grounds both the subjective character of experience and the value of respecting the autonomy of individuals.[5] Given the unique nature of personal experience, it is impossible to know objectively what it would be like to be another human being. However, Atkins argues that we can imagine not just what it would be like for me to be in the position of another human being in a specific set of circumstances, but also what it would be like to be that other human being in that same set of circumstances.[6] On this basis it might be possible to base a decision not to disclose genetic information on what we think the particular individual might want given our perception of her and her experiences. This substituted judgment protects her interest in autonomy because it is what she would want if she could make the choice herself.[7]

It is easier to appeal to the principle of autonomy to ground a subjective, as opposed to an objective, substituted judgment.[8] The problem with a subjective simulated judgment, which purports to further a person's wishes based on another's perception of her experiences, is that it requires knowledge of the individual and how her experiences and perceptions might influence her decision whether or not to know genetic information.[9] The health professional,

2   J. Raikka, 'Freedom and a Right (Not) to Know' 12 (1998), *Bioethics* 49, 60; J. Husted, 'Autonomy and a Right Not to Know' in R. Chadwick, M. Levitt and D. Schickle (eds), *The Right to Know and the Right Not to Know,* Aldershot: Avebury (1997), 55, 67.
3   See Chapter 4 for an in depth discussion of the doctrine of substituted judgment. To avoid repetition, the theoretical basis of the doctrine is not revisited here.
4   K. Atkins, 'Autonomy and the Subjective Character of Experience' 17 (2000), *Journal of Applied Philosophy* 71, 73.
5   Ibid.
6   Ibid. 75.
7   See the discussion of the doctrine of substituted judgment in Chapter 4.
8   See the deeper discussion of this issue in Chapter 4.
9   Some commentators believe that the concept of substituted judgment cannot be based on autonomy at all. See, for example, A. M. Torke, G. C. Alexander; J. Lantos, 'Substituted Judgment: the Limitations of Autonomy in Surrogate Decision Making' 23 (2008), *Journal of General Internal Medicine* 1514; J. F. Childress, 'The Place of Autonomy in Bioethics' 20 (1990), *The Hastings Center Report* 12, 15; T. G. Gutheil and P. S. Appelbaum, 'Substituted Judgment: Best Interests in Disguise' 13 (1983), *The Hastings Center Report* 8; R. Dworkin, 'Medical Law and Ethics in the Post-Autonomy Age' 68 (1993), *Indiana Law Journal* 727.

who is envisaged here as the person on whom the duty not to interfere with
autonomy would be imposed, may never have met the potential recipient and,
thus, may not have the information relevant to protect autonomy via substituted judgment.[10]

However, as argued in Chapter 4, the doctrine of substituted judgment can
also be construed objectively.[11] In the context of the lack of an explicit indication as to what an individual does want, any protection of what she would want
will be based on objective considerations of what people in her situation would
generally want to know. From this perspective, the interest in not knowing
would be based on a conception of rational autonomy rather than a liberal individualistic interpretation of the concept. On this basis, where an individual is
given information concerning her genetic risks and, upon receipt of that
information, she decides she would rather not have known about it, the question of whether there has been an interference with her autonomy would depend
on whether it is deemed to be rational not to want to receive the information.
The problem is then how to determine what it is rational to want, and not
want, to know? In Chapters 3, 4 and 6 it is argued that English negligence
law's perception of whether the claimant's reaction to the perceived wrong is
rational might be determined by reference to the court's perception of the ordinary person's reaction to that perceived wrong.

When this method of evaluating the rationality of the claimant's position
is considered in Chapter 4, it is argued that the rationality of the claimant's
position falls to be considered in the context of a clear rational housing,
which determines the limited range of circumstances in which a wrong
could arise, thereby demarcating the circumstances in which a person could
even raise a legal argument that her autonomy has been interfered with
because of the careless frustration of a decision made on her behalf. From the
perspective of rationality, the genomic claim considered in Chapter 4 which
rested on the individual's desire not to have been born, is different to the
claim based on the interest in not knowing because in Chapter 4 the claim
arose from a recognized wrong. It is easier to make an appeal to the concept
of autonomy to argue that people would not want to be interfered with in
ways which are objectively recognized as wrongs than in ways which do not
have status as such objective wrongs. Mullender argues that the pursuit of
negligence law's protective purpose might generally be associated with a
commitment to the ideal of autonomy.[12] In the context of the claim

---

10   In this book the question of disclosure is considered from the perspective of the imposition of duties
     on health professionals who may not have had any previous contact with the potential recipient.
11   See, in particular, J. A. Robertson, 'Organ Donations by Incompetents and the Substituted Judgment
     Doctrine' 76 (1976), *Columbia Law Review* 48; *Re AC* (1990) 573 A 2d 1235 (DC CA), Terry JA.
12   R. Mullender, 'English Negligence Law as a Human Practice' 21 (2009), *Law and Literature* 321,
     324–325. This argument relates to recognized negligence claims where the question of whether the
     particular action amounts to a wrong is settled.

considered in Chapter 4, the HFEA has carefully considered and stipulated when a decision that a future person should not be born is legitimate and should, therefore, be respected. Where the regulatory body charged with determining the legitimacy of a desire to screen out embryos deems a particular desire legitimate, where the action which gives effect to the legitimate desire is carelessly frustrated, a wrong occurs. A prior decision has been made, not necessarily by the person who is ultimately harmed,[13] but which indicates the rationality of the individual's view that she has been harmed.[14] However, where the interest in not knowing one's genetic information is concerned, there is no pre-existing wrong upon which we can begin to demarcate when a desire not to have known information might be considered rational and, therefore, autonomous. Where the action in question does not frustrate an express decision, there is no clear wrong in terms of a breach of relevant regulatory provisions or established professional practices, which indicates the undesirability of the action which creates the grievance. Thus, there is little scope for appealing to the principle of rational autonomy to provide a basis for the interest in not knowing unknown genomic unknowns.

The adoption of a concept of rational autonomy based on a benchmark of what the majority of ordinary people would think might be workable if there is a clear indication of what people generally consider to be rational.[15] With regard to the issue of knowing or not knowing genetic information about oneself, we are unlikely to find a clear consensus in any given set of circumstances. There is no clear objective framework which establishes what people might and might not want to know upon which an evaluation of rationality as conceived here could be made. The question of whether to have or not to have genetic information about oneself is not a decision which could so clearly elicit a consensual response from the majority of people. Although it might be possible to argue that the majority of people would want to have genetic information which might allow them a significant chance to avoid or delay the onset of illness,[16] it cannot similarly be argued that most people would not want to know when there is nothing or little that they can do in terms of prevention or delay.

13 Both the created child and the parents have arguably been harmed in terms of the interference with their autonomy in these circumstances; see Chapters 4 and 5 respectively.
14 See the central list of conditions which the HFEA warrants as sufficiently serious to justify PGD. Available HTTP: www.hfea.gov.uk/pgd-screening.html (accessed 11 May 2010).
15 For a deeper discussion of the plausibility of this position see Chapter 3.
16 For example, the uptake rate for predictive genetic testing for familial adenomatous polyposis, for which there is effective treatment, is around 80 per cent. D. G. R. Evans, E. R. Maher, R. Maclead, D. R. Davies and D. Craufurd, 'Uptake of Genetic Testing for Cancer Predisposition – Ethical Issues' 34 (1997), *Journal of Medical Genetics* 746.

In general, evidence suggests that the uptake of screening is lower when there is no treatment for a particular condition, but the number of people who access such testing is not insignificant.[17] Whilst some people might be averse to having information concerning risks that they can do nothing about, others may be eager to know all they can about themselves so that they can prepare for likely consequences and live life in the light of their likely future. Whilst the uptake of predictive testing for Huntington's disease, for which there is currently no treatment, is relatively low[18] (around 5–25 per cent in the UK and elsewhere),[19] it is not insignificant enough to assume a wholesale desire not to know. Furthermore, there are many other variables which might affect the rate of uptake of testing for a particular condition. If the particular treatment is painful, experimental or does not have a particularly high success rate, this might further impinge on whether a person wants to know about her genetic risk. Indeed, evidence demonstrates an uptake rate of around 50 per cent in testing for many conditions which can be managed but not cured.[20] This very brief survey demonstrates that it would be virtually impossible to meaningfully determine whether a preference not to have genetic information is rational if the question of rationality is determined by appealing to the perceived views of the majority of ordinary people. In the absence of a clear framework from which it can be deduced when it would be rational not to want genetic information about oneself, the interest in rational autonomy cannot be the basis of a general duty not to disclose genetic information.[21]

If there is no prevailing position as to whether a desire not to know is deemed to be rational, we cannot formulate general duties of care which can

17 The data on the uptake of predictive testing might not present an accurate picture of whether people would welcome an unsolicited disclosure of genetic information. However, there is a dearth of data regarding unsuspecting individual's views on unsolicited disclosure, whilst there is fairly significant data on the uptake of predictive testing, which can be relied on to give a good indication of what people want to know.

18 When compared to conditions for which there is some treatment or avoidance strategy.

19 M. R. Hayden, 'Predictive Testing for Huntington's disease: the Calm After the Storm' 356 (2000), *The Lancet* 1944. As Hayden acknowledges, this was significantly less than the 70–80 per cent uptake that had been predicted before testing was available.

20 This rate of uptake provides little assistance in determining what the majority of ordinary people might want to know, thereby providing little assistance in pouring content into the concept of substantive rationality as conceived here. See, for example, I. Christiaans, E. Birnie, G. J. Bonsel, A. A. M. Wilde and I. M. van Langen, 'Uptake of Genetic Counselling and Predictive DNA Testing in Hypertrophic Cardiomyopathy' 16 (2008), *European Journal of Human Genetics* 1201. This study demonstrated that the uptake for predictive genetic testing for hypertrophic cardiomyopathy, after detection of the causal mutation in the proband, was 39 per cent. Prevention of sudden cardiac death in patients with a high risk by means of an implantable cardioverter defibrillator is effective in this condition; M. E. Ropka, J. Wenzel, E. K. Phillips, M. Siadaty and J. T. Philbrick, 'Uptake Rates for Breast Cancer Genetic Testing: A Systematic Review' 15 (2006), *Cancer Epidemiology, Biomarkers & Prevention* 840, where a systematic review of 40 studies revealed a real uptake rate of 59 per cent of various forms of predictive testing for breast cancer.

21 This is not to say that other ethical concepts could not be the basis of a general duty not to disclose.

be justified because they are based on rationality. If there is a duty not to interfere with a person's autonomy, it has to be clear what interferes with her autonomy at the time of the disclosure which she subsequently evaluates as being contrary to her wishes. This could be evidenced by her explicitly stated opinion. However, where this is not the case because she did not explicitly stipulate her desires prior to the disclosure, there would have to be a clear reason why the proposed disclosure should have been evaluated as likely to be contrary to her desires and, therefore, an interference with her autonomy. Otherwise the imposition of a duty of care would expose health professionals to liability on the basis of unpredictable individual evaluations of the desirability of the disclosure. Where there is no clear view of what it is rational to want to know and not to know, disclosure cannot be a wrong which interferes with autonomy, even rational autonomy, because it was not clear at the time of the disclosure that protection of autonomy required non-disclosure.[22]

Conflict between the interest in not knowing unknown unknowns and knowing unknown unknowns

Where it is not known what people want to know in the face of existing relevant genetic information, the interest in not knowing and the interest in knowing conflict. Practically speaking, there cannot be a co-existing interest in knowing and an interest in not knowing, both of which are based on the principle of autonomy. In the face of not knowing what people want to know, if a duty of care were imposed which sought to protect the interest in autonomy, it would have to be based on an objective position which would seek to maximize autonomy. On this basis, it might be argued that the interest in knowing has a greater propensity to provide protection for the essential elements of autonomy than the interest in not knowing.

Many commentators argue that autonomy is restricted where there is a lack of relevant information because a crucial aspect of autonomy is the possession of information which enables meaningful self-determination.[23] According to Aristotle, 'all men desire to know'.[24] Harris and Keywood argue that there is a normal and reasonable presumption of a relationship between information and

22 In this way, the claim of interference with autonomy which is based on the evaluation of a simulated decision is based on a theory of rational autonomy and can only amount to an interference where there is evidence as to what amounts to rational.

23 J. Harris and K. Keywood, 'Ignorance, Information and Autonomy' 22 (2001), *Theoretical Medicine* 415; R. Rhodes, 'Genetic Links, Family Ties, and Social Bonds: Rights and Responsibilities in the Face of Genetic Knowledge' 23 (1998), *Journal of Medicine and Philosophy* 10. But for an account of how autonomy is not necessarily restricted by a lack of information, see, J. S. Taylor, 'Autonomy and Informed Consent: A Much Misunderstood Relationship' 38 (2004), *The Journal of Value Inquiry* 383; L. O. Ursin, 'Personal Autonomy and Informed Consent 12 (2009), *Medicine, Healthcare and Philosophy* 17; G. Dworkin, 'Autonomy and Behaviour Control' 6 (1976), *Hastings Center Report* 23.

24 W. D. Ross (ed.), *Aristotelis Topica et Sophistici Elenchii*, Oxford: Clarendon Press (1958), 1.

autonomy.[25] Others agree that relevant information is crucial in forming autonomous desires and making autonomous choices.[26] On this basis, the principle of autonomy requires the provision of information which will facilitate meaningful decision-making.

In the legal context, autonomy finds its expression through the law of consent,[27] where the emphasis is firmly on the provision of relevant information.[28] Indeed, consent is vitiated where it is ill-informed.[29] The legal doctrine of informed consent is particularly well developed in relation to medical treatment. Here, failures to provide adequate information regarding the risks of, or the nature of, the treatment are actionable in negligence and battery respectively. Nowadays it might be argued that doctors do not generally withhold relevant information from patients when the patient is being asked to consent to a particular intervention. The doctrine of therapeutic privilege traditionally allowed doctors to discretionally withhold information from patients where it was thought that the disclosure would harm the patient. However, recent General Medical Council (GMC) Guidance suggests that the application of this doctrine in modern medicine is confined to very narrow circumstances where disclosure would cause the patient 'serious harm', which excludes the patient becoming upset or refusing treatment.[30] Thus, the modern assumption is that competent patients want, and can cope with, the relevant information and that it should be disclosed to allow them to form autonomous choices and desires.[31] If autonomy justifies the interest

25 J. Harris and K. Keywood, 'Ignorance, Information and Autonomy' 22 (2001), *Theoretical Medicine* 415, 417.

26 See, for example, R. Rhodes, 'Genetic Links, Family Ties, and Social Bonds: Rights and Responsibilities in the Face of Genetic Knowledge' 23 (1998), *Journal of Medicine and Philosophy* 10, 17–18; G. Laurie, 'The Most Personal Information of All: An Appraisal of Genetic Privacy in the Shadow of the Human Genome Project' 10 (1996), *International Journal of Law, Policy and the Family* 74, 87; G. Laurie, 'Protecting and Promoting Privacy in an Uncertain World: Further Defences of Ignorance and the Right not to Know' (2000) 7, *European Journal of Health Law* 185, 189.

27 J. Harris and K. Keywood, 'Ignorance, Information and Autonomy' 22 (2001), *Theoretical Medicine* 415, 417.

28 Ibid.

29 J. Harris, *The Value of Life*, London: Routledge (1985), Chapter 10.

30 General Medical Council, *Consent, Patients and Doctors Making Decisions Together*, London (2008), paragraph 16.

31 Although historically there might have been a tendency on behalf of medical profession to withhold information from patients on the basis that the patient might be too naïve or sensitive to cope with that information, this position is now generally accepted as being unduly paternalistic. There may be other sound ethical reasons for a general duty not to disclose, relating to prevention of harm or the promotion of benefit as opposed to the protection of autonomy. Indeed, in the medical context where it has been accepted that information should be withheld from patients, the rationale for doing so has been seen in terms of the obligation of doctors not to harm their patients rather than any entitlement that patients might have to shield themselves from unpleasant truths. See J. Harris and K. Keywood, 'Ignorance, Information and Autonomy' 22 (2001) *Theoretical Medicine* 415, 415.

in knowing, it cannot also justify the interest in not knowing.[32] Thus, although a person subsequently interprets her knowledge of genetic information about herself as deleterious and concludes that she would rather not have had that information, the disclosure cannot amount to an interference with her autonomy because autonomy is protected through the provision of relevant information. That is, where the two interests potentially conflict, because it is not known what the person wants to know, the interest in knowing more clearly protects one of the core aspects of autonomy, that of information provision, than the interest in not knowing.

Autonomy and the interest in not knowing a known unknown

Although it has been argued that the interest in not knowing an *unknown* unknown cannot easily be justified by recourse to the principle of autonomy, particularly in the face of the competing autonomy interest in knowing an *unknown* unknown, a different argument can be made from the perspective of autonomy for the interest in not knowing *known* unknowns.[33] Assume an individual has stipulated, in a way readily discoverable by health professionals, that she does not want to know information about her genetic risks. However, ignoring this direction, a health professional informs the individual of her specific risk which the professional is aware of because of a genetic test performed on a relative. Here there is a prior choice not to be informed, which represents the individual's wishes. Because the prior wish is readily identifiable by the health professional, rather than based on a subsequent reaction to the disclosure of information, it is not unduly onerous to expect the health professional to respect that wish.

However, even where the individual has expressly registered her desire not to receive genetic information, it might be argued that that desire cannot be justified by reference to autonomy because information is so

---

32 R. Rhodes, 'Genetic Links, Family Ties, and Social Bonds: Rights and Responsibilities in the Face of Genetic Knowledge' 23 (1998), *Journal of Medicine and Philosophy* 10, 18. For agreement see; J. Harris and K. Keywood, 'Ignorance, Information and Autonomy' 22 (2001), *Theoretical Medicine* 415.

33 One commentator who has given this area significant consideration agrees that whilst it is difficult to justify the interest in not knowing unknown unknowns by recourse to personal autonomy, the interest in not knowing known unknowns can be justified on this basis. See G. Laurie, 'The Most Personal Information of All: An Appraisal of Genetic Privacy in the Shadow of the Human Genome Project' 10 (1006), *International Journal of Law, Policy and the Family* 74; G. Laurie, 'Obligations Arising from Genetic Information – Negligence and the Protection of Familial Interests' 11 (1999), *Child and Family Law Quarterly*; G. Laurie, 'Genetics and Patients' Rights: Where are the Limits?' 5 (2000), *Medical Law International* 25; G. Laurie, Protecting and Promoting Privacy in an Uncertain World: Further Defences of Ignorance and the Right Not to Know' 7 (2000), *European Journal of Health Law* 185; G. Laurie, *Genetic Privacy: A Challenge to Medico-Legal Norms*, Cambridge: Cambridge University Press (2002).

central to autonomy that a refusal of information can never be justified by an appeal to the principle of autonomy. Where the decision not to know refers to specific genetic information, the individual knows that there is information to know and the nature of that information, thus in essence she knows what it is that remains unknown. Nevertheless, even in the face of such an express stipulation, some commentators argue that a choice not to receive relevant genetic or medical information about oneself cannot be defended by reference to claims to autonomy.[34] Thus, the health professional does not interfere with the individual's autonomy when she discloses information, even if this is contrary to the recipient's expressed wishes. If this is the case, only acceptances of relevant information could be seen as autonomous; refusals could not be so construed. This position is similar to that argued above, in relation to not knowing unknown unknowns, whereby information is crucial to autonomy and therefore no justification for refusing information, can be found by appealing to autonomy.[35] It might be argued that those who take this position are subscribing to a conception of rational autonomy. That is, rather than arguing that information cannot be autonomously refused per se, the argument is that an appeal cannot be made to autonomy to refuse information that you ought rationally to know. Harris and Keywood state:

> For where the individual is ignorant of information that bears upon rational life choices she is not in a position to be self-governing. If I lack information, for example about how long my life is likely to continue I cannot make rational plans for the rest of my life.[36]

This seems to suggest that autonomous decisions are necessarily rational decisions and that rational decisions cannot be made without the relevant information.

The objective assumption is that people cannot rationally refuse any relevant information about themselves and the implication is that, if they do, they are not rationally autonomous.[37] On this basis, a disclosure of genetic information to an individual against her will does not interfere with her autonomy because her refusal to receive the information was not rational and, therefore, autonomous. This perspective presents a substantive account of rational autonomy, whereby choices must accord with some objective notion of value, namely the acceptance of information which is relevant to rational choices.

---

34 See, in particular, J. Harris and K. Keywood, 'Ignorance, Information and Autonomy' 22 (2001), *Theoretical Medicine* 415; R. Rhodes, 'Genetic Links, Family Ties, and Social Bonds: Rights and Responsibilities in the Face of Genetic Knowledge' 23 (1998), *Journal of Medicine and Philosophy* 10.

35 J. Harris and K. Keywood, 'Ignorance, Information and Autonomy' 22 (2001), *Theoretical Medicine* 415, 419.

36 Ibid. 421.

37 And therefore your choice need not be respected on the basis of autonomy.

Harris and Keywood believe that this objective notion of value rationality, as the universal receipt of information about oneself, stems from the principle of autonomy itself as a principle which fundamentally protects the giving of information. However, others do not believe that the receipt of information is always required for a person to act autonomously.[38] For these theorists it is not inimical to autonomy to authentically and independently refuse information. Indeed, even for Harris and Keywood the problem with justifying a refusal of information by recourse to autonomy is not that decisions which foreclose future autonomy cannot be based on autonomy,[39] but rather that when adopting a conception of autonomy as involving the exercise of control, as they do,[40] ignorance of crucial information is inimical to autonomy because it interferes with the ability to control one's own destiny.

Rhodes adopts a similar position whereby the interest in not knowing cannot be based on autonomy where the information is relevant to making rational decisions. However, the value of rationality she employs does not seem to arise directly from the concept of autonomy as centred on the knowledge of relevant information, but from the perspective of the reasonable person and the information which she would want to know:

> I am obligated to make thoughtful and informed decisions without being swayed by irrational emotions including my fear of knowing significant genetic facts about myself. When I recognize that I am ethically required to be autonomous, I must also see that since autonomous action requires being informed of what a reasonable person would want to know under the circumstance, I am ethically required to be informed.

The adoption of a notion of value, based on what the majority of ordinary people would want, reflects the argument throughout this book that if the English courts were to deem that rationality is an aspect of autonomy, the concept of rationality adopted might be based on the court's perception of what the majority of ordinary people would consider to be rational.[41]

38 R. Andorno, 'The Right Not to Know: An Autonomy Based Approach' 30 (2004), *Journal of Medical Ethics* 435; J. Raikka, 'Freedom and a Right (Not) to Know' 12 (1998), *Bioethics* 49, 60;. J. Husted, 'Autonomy and a Right Not to Know' in R. Chadwick, M. Levitt and D. Schickle (eds), *The Right to Know and the Right Not to Know,* Aldershot: Avebury (1997), 55, 67; G. Dworkin, 'Autonomy and Behaviour Control' 6 (1976); *Hastings Center Report* 23; T. Takala, 'The Right to Genetic Ignorance Confirmed' 13 (1999), *Bioethics* 288; L. O. Ursin, 'Personal Autonomy and Informed Consent' 12 (2009) *Medicine Healthcare and Philosophy* 17.

39 Indeed, they argue that suicide and euthanasia are consistent with the idea of autonomy.

40 J. Harris and K. Keywood, 'Ignorance, Information and Autonomy' 22 (2001), *Theoretical Medicine* 415, 420.

41 On the basis that the House of Lords has frequently relied on this notion of value to determine whether the law should recognize harm. See the in depth discussion of this issue in Chapter 3. It also relies on this notion with respect to the setting of the standard of care in negligence.

However, the problem with applying such a notion of value rationality to the question of whether it is rational to refuse genetic information about oneself is that, as noted above in the scenario concerning subsequent reactions to unknown unknowns, it will be difficult to form a conclusion about when the majority of ordinary people would deem it rational to not want to know personal genetic risks. Whilst many might perceive a refusal of information concerning a genetic condition which can be neither prevented nor delayed to be rational, would they consider it rational to refuse information where there is treatment, but it is experimental or particularly invasive, painful and frequent? An analysis of when it might be substantively rational to refuse information regarding one's genetic risks is difficult where substance is poured into the notion of rationality by reference to ordinary people and there is no clear evidence of the views of ordinary people. Given this, let us focus on rational autonomy as procedurally construed, and consider the interest in not knowing known unknowns from the perspective of autonomy as a procedural concept.

The interest in not knowing known unknowns from the perspective of procedurally rational autonomy

From the value rationality perspective just discussed, the value of a decision rests on its specific content: information is valuable to rational decision-making, so a decision to refuse information cannot be rational and, therefore, autonomous. However, conceptions of autonomy that rest on procedure do not require decisions to have specific content. Procedurally speaking, an autonomous individual or decision must be independent in the sense of freedom from (illegitimate) interference, but the content of the decision need not meet with objective criteria. Gerald Dworkin has produced possibly the most prevalent and influential theory of the concept of autonomy as procedural independence from others. According to Dworkin, autonomy is characterized by the formula; authenticity + independence = autonomy.[42] In essence, this means that the autonomous person is one who does *his own* thing.[43] This requires an analysis of what it is for a motivation to be *his* and what it is for it to be his *own*.[44] Dworkin calls the former authenticity and the latter independence.[45] Other commentators who adopt a procedural interpretation of autonomy agree that authenticity and independence are

42   G. Dworkin, 'Autonomy and Behaviour Control' 6 (1976), *Hastings Center Report* 23, 24; G. Dworkin, *The Theory and Practice of Autonomy*, Cambridge: Cambridge University Press (1988), Chapter 1.

43   G. Dworkin, 'Autonomy and Behaviour Control' 6 (1976), *Hastings Center Report* 23, 24 (original emphasis).

44   Ibid.

45   Ibid.

crucial elements.[46] To be authentic the individual must possess certain abilities.[47] By distinguishing between first and second order desires, some autonomy theorists define authenticity as the ability to raise the question of whether one identifies with the motivations for one's actions.[48] For example, one may be motivated by jealousy or anger but desire that one's motivations be different.[49] In other words, people are capable of wanting to be different in their preferences and purposes from what they actually are.[50] On this view, autonomy is related to the individual's ability to critically analyse her first order motivations and to change them if she wishes. From this perspective, many types of interference might hinder autonomy because they affect the individual's ability to either make or reject identifications at the second order with her first order motivations.[51]

Dworkin argues that whilst authenticity is necessary for autonomy, it is not sufficient because a person's motivational structure might be hers without being her own.[52] This may be because the identification with her motivations has been influenced in a decisive way by others. Dworkin calls this a lack of procedural independence.[53] It does not automatically follow that any interference with an individual's motivational structure prevents procedural independence with the result that the individual's decisions cannot be considered to be her own. Indeed, Dworkin devotes significant attention to distinguishing between those interferences which prevent the individual's decision from being her own and those which do not:

> With respect to autonomy conceived of as authenticity under conditions of procedural independence, the paradigms of interference are manipulation and deception, and the analytic task is to distinguish these ways of

46 See, in particular, J. Feinberg, 'Autonomy' in *The Inner Citadel: Essays on Individual Autonomy*, J. Christman (ed.), Oxford: Oxford University Press (1989); R. Young, 'Autonomy and The Inner Self' in *The Inner Citadel: Essays on Individual Autonomy*, J. Christman (ed.), Oxford: Oxford University Press (1989), 79.

47 The issue of authenticity with respect to the refusal to receive genetic information is considered in detail below.

48 See, in particular, G. Dworkin, *The Theory and Practice of Autonomy*, Cambridge: Cambridge University Press (1988) 15; F. Frankfurt, *The Importance of What We Care About: Philosophical Essay*, Cambridge: Cambridge University Press (1988); J. Feinberg, Autonomy in *The Inner Citadel: Individual Essays on Personal Autonomy*, J. Christman (ed.), Oxford: Oxford University Press (1989), 36–38 on moral authenticity.

49 G. Dworkin, *The Theory and Practice of Autonomy*, Cambridge: Cambridge University Press (1988), 15.

50 H. Frankfurt, 'Freedom of the Will and the Concept of a Person' in *The Inner Citadel: Individual Essays on Personal Autonomy*, J Christman (ed.), Oxford: Oxford University Press (1989), 64.

51 G. Dworkin, *The Theory and Practice of Autonomy*, Cambridge: Cambridge University Press (1988), 15–17.

52 G. Dworkin, 'Autonomy and Behaviour Control' 6 (1976), *Hastings Center Report* 23, 25.

53 Ibid. 25.

influencing people's higher order judgements from those (education, requirements of logical thinking, provision of role models) which do not negate procedural independence.[54]

If authenticity concerns the ability to reflect critically on first order motivations, methods which prevent such reflection will prevent procedural independence. Dworkin argues that it is a feature of persons that they are able to reflect on their decisions, motives, desires, habits and so forth,[55] suggesting that this ability is largely innate in humans. Thus, some ways of influencing people's higher order judgments, such as education and the provision of role models, do not negate procedural independence because people will retain the ability to reflect on their motivations. The paradigm cases of influences with procedural independence destroy the ability of the agent to reflect critically on her motivations, rather than simply influence those motivations without interfering with the ability to reflect critically upon them. From this perspective, methods of influencing motivation of which the individual is not conscious might interfere with her procedural independence because she is not aware of the true determinants of her behaviour.[56] In essence, this is a reflection of the notion that authentic behaviour leaves no room for false consciousness.[57] Methods which make the psychic costs of the reflective enterprise so painful that coercion or manipulation occur also negate procedural independence.[58] According to Dworkin, methods which interfere with procedural independence, thereby negating autonomy in the sense of keeping the individual ignorant of the determinants of her behaviour, might be subliminal motivation or the destruction of parts of the brain necessary for performing the critical reflection which authenticity requires. In these instances, since the individual does not know the real reasons for her actions, she cannot reflect on those reasons and make a favourable or adverse judgement concerning them.[59] On the other hand, manipulative methods such as threats and physical force might not interfere with the knowledge that a particular determinant is influencing behaviour, but simply the ability to critically reflect on that influence.

Related to the first type of interference, Dworkin argues that methods which rely on deception, on keeping the individual in ignorance of relevant facts, can interfere with procedural independence.[60] From this perspective, it seems that the failure to disclose relevant genetic information considered in

---

54 Ibid. 26.
55 G. Dworkin, 'Autonomy and Behaviour Control' 6 (1976), *Hastings Center Report* 23, 24; G. Dworkin, *The Theory and Practice of Autonomy*, Cambridge: Cambridge University Press (1988), 15.
56 G. Dworkin, 'Autonomy and Behaviour Control' 6 (1976), *Hastings Center Report* 23, 26.
57 Ibid. 25.
58 Ibid. 26.
59 Ibid.
60 Ibid. 27.

the previous chapter would violate the conception of procedural autonomy conceived as authenticity + procedural independence. Given this, an appeal to this interpretation of procedural autonomy does not provide support for the interest in not knowing unknown unknowns considered above. However, Dworkin does recognize that a person who authentically wishes that her procedural independence be restricted in certain ways, can act in a procedurally independent manner in renouncing her procedural independence, such that she has acted authentically and thereby autonomously restricted her autonomy.[61]

According to Dworkin:

> a person might decide to renounce his independence of action or thought because he wants (genuinely) to be that sort of person. A person might do whatever his mother, or his government, tells him to do, and do so in a procedurally independent manner.[62]

He continues, the person who 'wishes to be restricted in various ways, whether by the discipline of the monastery, regimentation of the army, or even by coercion, is not, on that account alone, less autonomous.'[63] Thus, where one authentically, in the sense that she has the ability to raise the question of whether she identifies with the motivations for her actions and does identify with her motivations, and procedurally independently wishes to renounce future independence, that decision cannot be criticized on the basis that it is not procedurally independent. Given that this is how the person wants to be motivated, she has authentically and independently restricted her future independence of action or thought in an autonomous way. Procedural independence as conceived here, relates to an inability to critically reflect on one's motivational structure. In the scenario outlined above, the person is able to reflect, which she has done, and has made an authentic decision about how she wants her life to be.

---

61 However, others maintain that it is crucial to a procedural account of autonomy that the individual retain control over her decisions and actions. They argue that acts and decisions which appear to be independent but forgo future independence cannot be justified by recourse to the principle of autonomy. On this basis, one cannot defer independent judgment whether to government, moral authority, God or another mere mortal on the basis of autonomy. See, for example, R. Wolff, *In Defense of Anarchism*, New York: Harper & Row (1970), 41; J. Rachels, 'God and Human Attitudes' 7 (1971), *Religious Studies* 334; M. Osiel, *Obeying Orders, Atrocity, Military Discipline and the Law of War*, New Jersey: Transaction (1999); R. Rhodes, 'Genetic Links, Family Ties, and Social Bonds: Rights and Responsibilities in the Face of Genetic Knowledge' 23 (1998), *Journal of Medicine and Philosophy* 10. It might be argued that this position does not describe procedural autonomy at all, but a substantive concept of autonomy where a decision whose content is to renounce future independence is not autonomous because of an error in the substance of the decision rather than in the procedure.
62 G. Dworkin, 'Autonomy and Behaviour Control' 6 (1976), *Hastings Center Report* 23, 25; G. Dworkin, *The Theory and Practice of Autonomy*, Cambridge: Cambridge University Press (1988), Chapter 2.
63 G. Dworkin, *The Theory and Practice of Autonomy*, Cambridge: Cambridge University Press (1988), 18.

Dworkin maintains that all choices to some extent foreclose other choices, reversibly or irreversibly, and such foreclosures need not be viewed as forfeitures of autonomy.[64] This position is possible from Dworkin's conception of autonomy as resting on procedural independence, as opposed to substantive independence. Substantive independence demands that the individual not defer independent judgment. In Dworkin's opinion, if this were a feature of autonomy it would make autonomy inconsistent with loyalty, objectivity, commitment, benevolence and love.[65] Consequently, he argues that a conception of autonomy that insists on substantive independence should not have claim to our respect as an ideal.[66] According to Dworkin, what is an important aspect of autonomy as procedurally conceived 'is that the commitments and promises a person makes be ones he views as his, as part of the person he wants to be, so that he defines himself via those commitments'.[67]

As the above discussion demonstrates, some commentators do not accept the argument that one can appeal to the principle of autonomy, however conceived, to argue that a knowledgeable refusal of relevant genetic information ought to be respected. However, two prominent advocates of this view concede that there can be legitimate cases of autonomously chosen restrictions on autonomy.[68] However, they argue that a distinction must be made between autonomously chosen restrictions on autonomy which are consistent with autonomy, understood as an ethical principle, and such choices which are inconsistent with autonomy.[69] Harris and Keywood contrast the monk who enters the monastic order, but remains free to leave as an autonomously chosen restriction on autonomy, and the person who sells herself into slavery as a choice which is inconsistent with the idea of autonomy, and cannot be protected by appeals to autonomy as a moral principle.[70] The monk remains fully autonomous because although agreeing to be bound, he is still free to choose.[71] As we know, Dworkin refutes this position because he believes that what is valuable about autonomy is that the commitments a person makes are ones she views as hers, as part of the person she wants to be, so that she defines herself via those commitments. But whether they are long or short term, prima facie or absolute, permanent or temporary, is not what contributes to their value.[72]

64 Ibid. 26.
65 Ibid. 21.
66 Ibid. 21 and 25.
67 Ibid. 26.
68 J. Harris and K. Keywood, 'Ignorance, Information and Autonomy' 22 (2001), *Theoretical Medicine* 415, 419.
69 Ibid. 419.
70 Ibid. 420.
71 Ibid. 419.
72 G. Dworkin, *The Theory and Practice of Autonomy*, Cambridge: Cambridge University Press (1988), 26. Contrast this view with that of Harris and Keywood as to when it is possible to autonomously foreclose independence in a way which is consistent with autonomy above and at 419–420.

If Harris and Keywood's conception is correct, it is not clear why the choice not to accept genetic information about oneself is akin to the example of the slave rather than the monk.[73] A person who chooses not to receive information is not bound by her original decision so that she cannot subsequently ask for the information when and if she wants it, or when she approaches a decision where she feels genetic information might be relevant. Indeed, she can always choose to have genetic tests *herself* to discover more accurate information about her own genetic constitution than that which she might discover as a result of a relative's test.

The appeal to autonomy to ground the specific interest in not knowing genetic information has been endorsed by many commentators.[74] The interest has also been recognized by two international instruments which are credited with playing a crucial role in establishing rights in the genetic context.[75] Article 10(2) of the Council of Europe Convention on Human Rights and Biomedicine states: 'Everyone is entitled to know any information collected about his or her health. However, the wishes of individuals not to be so informed shall be observed.' Similarly, Article 5(c) of the UNESCO Universal Declaration on the Human Genome and Human Rights states: 'The right of every individual to decide whether or not to be informed of the results of genetic examination and the resulting consequences should be respected.' Whilst neither of these instruments specifically connects the right not to know to the principle of autonomy, the foundation of these interests lies in autonomy as a means of protecting personal choice.

Dworkin specifically endorses the argument that one can renounce her independence (authentically) in a procedurally independent manner which can, thus, be justified by an appeal to autonomy with respect to the knowledgeable and expressed interest in not knowing medical information about oneself: 'If a patient has knowingly and freely requested of the doctor that he not be informed or consulted about his course of treatment then to seek to obtain informed consent would itself be a denial of autonomy.'[76] This reflects

73   As they argue that it does. J. Harris and K. Keywood, 'Ignorance, Information and Autonomy' 22 (2001), *Theoretical Medicine* 415, 419–421.

74   R. Andorno, 'The Right Not to Know: An Autonomy Based Approach' 30 (2004), *Journal of Medical Ethics* 435; J. Raikka, 'Freedom and a Right (Not) to Know' 12 (1998), *Bioethics* 49, 60; J. Husted, 'Autonomy and a Right Not to Know' in R. Chadwick, M. Levitt and D. Schickle (eds), *The Right to Know and the Right Not to Know*, Aldershot: Avebury (1997), 55, 67; G. Dworkin, 'Autonomy and Behaviour Control' 6 (1976), *Hastings Center Report* 23; T. Takala, 'The Right to Genetic Ignorance Confirmed' 13 (1999), *Bioethics* 288; L. O. Ursin, 'Personal Autonomy and Informed Consent' 12 (2009), *Medicine Healthcare and Philosophy* 17.

75   G. Laurie, 'Genetics and Patients' Rights: Where are the Limits?' 5 (2000), *Medical Law International* 25, 26.

76   G. Dworkin, *The Theory and Practice of Autonomy*, Cambridge: Cambridge University Press (1988), 118.

Dworkin's view that more knowledge is not always better than less because there are times when we really do not want to know.[77]

However, on a Dworkinian approach to procedural autonomy, procedural independence is not sufficient to demonstrate autonomy; authenticity is also necessary. People have many desires and motivations and, at times, these desires and motivations might come into conflict. Generally, people feel uncomfortable in the face of clear conflict in their desires and motivations and will, therefore, try to change one or the other to restore harmony.[78] Some believe that the voluntariness of a person's actions can be considered solely at the level of the promotion or hindrance of the desires which move her to action;[79] often termed first order desires.[80] However, other commentators argue that focusing on first order desires ignores a crucial feature of persons, which is that they are able to reflect upon and adopt attitudes to their first order desires, wishes, motives and habits.[81] Frankfurt argues that the capacity for self-reflection is a characteristic of humans which sets us apart from other animals.[82] Humans are able to reflect on their decisions, motives, desires and habits and in doing so form preferences concerning these.[83]

The essence of this theory is that besides wanting to do something and choosing to act upon that desire, people are capable of wanting to be different. The question of whether or not a person acts freely is determined by the attitudes a person takes towards the reasons upon which she acts. If she identifies with those reasons and assimilates them to herself, she acts freely.[84] If,

---

77 G. Dworkin, 'Autonomy and Behaviour Control' 6 (1976), *Hastings Center Report* 23, 27.

78 D. J. Koehler, 'Explanation, Imagination and Confidence in Judgment' 110 (1991), *Psychological Bulletin* 499, 508. Dworkin gives an example of how cognitive dissonance methods of changing behaviour might violate procedural accounts of autonomy by creating causal influences of which the individual is not conscious and, therefore, cannot reflect on. However, the experiment described also demonstrates the human desire to eliminate conflict in their desires and motivations. Children are asked to rank a number of toys in order of preference. The adult then leaves the room and warns them not to play with the most preferred toy. He then returns and the children are asked again to order their preferences. The children shift their first choice down the list of preference. Cognitive dissonance theory predicts this on the basis of a conflict between preference and action inconsistent with preference. The conflict is resolved by shifting preference.

79 See, for example, J. P. Plamenatz, *Consent, Freedom and Political Obligation*, Oxford: Oxford University Press (1938), 110; T. V. Daveney, 'Wanting' 11 (1961), *Philosophical Quarterly* 139.

80 See, in particular, G. Dworkin, *The Theory and Practice of Autonomy*, Cambridge: Cambridge University Press (1988) and H. Frankfurt, *The Importance of What We Care About: Philosophical Essays*, Cambridge: Cambridge University Press (1988).

81 Frankfurt and Dworkin are the two most notable advocates of this hierarchical approach to determining whether a person acts freely.

82 H. Frankfurt, 'Freedom of the Will and the Concept of a Person' in *The Inner Citadel: Individual Essays on Personal Autonomy*, J. Christman (ed.), Oxford: Oxford University Press (1989), 64.

83 G. Dworkin, 'Autonomy and Behaviour Control' 6 (1976), *Hastings Center Report* 23, 24; G. Dworkin, *The Theory and Practice of Autonomy*, Cambridge: Cambridge University Press (1988), 15.

84 G. Dworkin, 'Acting Freely' 4 (1970), *Nous* 367, 377.

however, she feels alienated from what influences her to act and prefers to be the kind of person who is motivated in different ways, her actions will not be viewed as her own.[85] Dworkin argues that such identification demonstrates authenticity, which is a necessary element of autonomy. Thus, autonomy is conceived as a second order capacity of persons to reflect critically on their first order preferences, desires, wishes and so forth and the capacity to accept or attempt to change these in light of higher order preferences and values.[86] Autonomy is therefore thought of as approval and integration at the highest order. Where there is conflict between a person's first order desire and her second order motivations, she could, in theory, choose to change or accept a desire at either the first or at the higher level to resolve the conflict in her desires, thereby identifying with her influences and assimilating them to herself so that they may be viewed as hers, and therefore described as authentic.

Assume an individual is motivated to perform a particular action by envy. Dworkin argues that 'one way of becoming autonomous is to cease to be motivated by envy. But another way ... is to change one's objections to envy, to change one's second order preferences'.[87] So, a person can accept her motivations or alter them, or alter her first order desires to make her preferences effective in action. However, where conflict remains between one's first order desires and her preferences and desires at a higher order, she will remain isolated from the desires which move her to action. To give an example;[88] a person may want to break the habit of smoking and prefer to stop smoking because she recognizes its harmful character. It might be that recognition alone is effective in changing her behaviour. However, as is well known, many people appear incapable of simply giving up smoking and may therefore remain alienated from the motivation which moves them to action because they would prefer that their higher motivation not to smoke were effective in action. However, there are other ways that a person might make her motivation authentically hers, although in practice these are not necessarily easier to institute than simply changing one's behaviour. If the causal path of simply changing her behaviour is closed,[89] she might prefer to have a different causal structure introduced which makes her feel sick upon smoking. Even though her behaviour is not then under her voluntary control, she may wish to be motivated in this way to stop smoking. She would then view the causal

---

85 G. Dworkin, *The Theory and Practice of Autonomy*, Cambridge: Cambridge University Press (1988), 15.

86 Ibid. 20.

87 Ibid. 16.

88 This example is drawn from Dworkin's work. See G. Dworkin, 'Autonomy and Behaviour Control' 6 (1976), *Hastings Center Report* 23, 24; G. Dworkin, *The Theory and Practice of Autonomy*, Cambridge: Cambridge University Press (1988), 15.

89 Because she cannot just stop smoking.

influences as hers because the part of her which wishes to stop smoking is the part of her she wants to see carried out and that which she considers to be her true self.[90]

If a procedural concept of autonomy, which requires that the first order desires which move a person to action are authentically hers before she is to be deemed to be acting voluntarily, were adopted with respect to the question of whether a decision not to know relevant genetic information is autonomous, we need a level of analysis that goes deeper than that required where autonomy is conceived as pertaining only to the promotion or hindrance of first order desires. First, to fulfil the authenticity condition, the individual expressing her wish not to know must be capable of raising the question of whether she will identify or reject the reason upon which she acts. She is moved by a desire not to know genetic information and may identify with that desire at a higher order, in that when she reflects upon the desire that moves her to action, she desires to have that desire. In other words, upon reflection, she reflects that she does not want to have that information and, therefore, identifies with the influences that motivate her to reject it. On the other hand, she might express a desire not to know, but when she reflects upon this first order desire, she realizes that it is motivated by fear. She may reflect that she does not want to be motivated by fear and wishes that she had the courage to know the genetic information. This higher order desire reflects the wishes she wants to see carried out and reflects her true self. She might resolve the authenticity-frustrating conflict by simply changing her behaviour so it is motivated by the desire to have relevant information, rather than by fear. Alternatively, but perhaps less realistically, she might change her second order preferences so that she no longer objects to being motivated by fear. However, it might be argued that the higher level desire here, the desire not to be motivated by fear, reflects deeply held values which could not be easily changed. From this perspective, where an individual expresses an interest in not knowing relevant genetic information, a deeper analysis of the motivation for her decision would be required to ascertain whether she identifies with the influences which motivate her to action. If she does not, her decision to refuse the relevant information does not possess the necessary conditions for autonomy and, from this procedural perspective, a failure to respect her decision will not amount to an interference with her autonomy. However, if she has the ability to critically reflect on her first order desire not to know, and she accepts it in the light of her higher

---

90 G. Dworkin, 'Autonomy and Behaviour Control' 6 (1976), *Hastings Center Report* 23, 24. This interpretation raises questions with regard to responsibility for action. For example, C. J. Moya and S. E. Cuypers, 'Responsibility for Action and Belief' 12 (2009), *Philosophical Explorations* 81. This issue is beyond the scope of this discussion.

order preference, the desire which moves her to action can be deemed autonomous and an appeal to respect this decision can be made in the name of autonomy.[91]

Conclusion

This discussion argues that a decision to refuse relevant genetic information about oneself can be justified via an appeal to a conception of autonomy conceived as a procedural, rather than substantive, concept. Indeed, in this scenario where the concept of rationality as referenced to the perspective of the majority of ordinary people cannot be easily established, imbuing the principle of autonomy with value rationality does not provide analytical purchase which might enable external evaluation of what might amount to an interference with autonomy. However, if the English tort system were specifically imbued with protection of the interest in autonomy as procedurally conceived, a person's stated interest in not knowing genetic information about herself could be justified by an appeal to the concept of autonomy. However, a negligence claim would require more than a harm. It would require fault on behalf of the health professional who disclosed the unwanted information. Whilst, as intimated above, the attribution of fault in the face of a disclosure of an unknown unknown raises too many practical difficulties for it to be feasible, where a person has specifically requested that she remain ignorant of a known unknown, there is significantly greater potential for attributing fault when she is made known of that unknown.[92]

91 If the interference with autonomy that might be occasioned by the disclosure of a known unknown in the face of an explicit refusal were recognized as a legal harm, this would raise the question of where the value lies in autonomy; that is, whether it is in the objects of choice of whether the value of autonomy is intrinsic and independent of what one wishes to bring about. In other words, adverse psychiatric consequences arising from an unwanted disclosure might be seen as deleterious from a legal perspective, whilst the disclosure in the face of the refusal without more, might not be seen as deleterious. As argued in Chapter 5, a true recognition of the value of autonomy would lead to recognition of the harm suffered in the latter as well as the former. See the discussion in Chapters 3 and 5 regarding the intrinsic and instrumental value of autonomy. To prevent repetition this issue is not also considered here.

92 The question of fault is at the heart of the issue of breach of duty. The discussion of the concept of the ordinary person in this chapter regarding the issue of what might be recognized as actionable damage may also have a role in determining breach of duty. Thank you to Professor Aurora Plomer for highlighting this point. As discussed in Chapter 2, this book focuses on the issue of legal damage and corresponding duties of care. This is not to suggest that there is not significant debate to be had regarding the issue of breach of duty with regard to this novel claim and, for that matter, all the novel claims considered in this book. However, there is not the space here to consider all the elements of negligence in relation to all the claims, hence the focus on damage and corresponding duties.

Conclusion

Modern genetic technology is changing what we know about health by enhancing our understanding of how genes function. Information and services generated by this technology enable the prediction and modification of biological futures. This changing social context might lead to new kinds of grievance, which could be articulated as novel legal challenges. In the absence of a contract or dedicated regulation, the tort system would provide the most likely avenue for those wanting to pursue claims based on grievances arising from genetic services. It is argued here that the suite of novel genomic claims considered in this book could be presented as novel challenges within the tort of negligence. This book considers how English negligence law might react to four genomic negligence claims arising from reproductive genetic services and the treatment of genetic information. First, the discussion focuses on the English courts' potential reaction to the genomic claims within the context of the current approach in English negligence law. I have shown that, given the courts conservative approach to the recognition of new types of damage and the imposition of corresponding duties of care, the novel genomic claims are largely likely to be filtered out at an early stage without extensive consideration, on the basis that the claimant has not suffered any harm which is recognized in the tort of negligence. One of the major problems for the claimant is that the occurrence, upon which her grievance is based, is not one which is universally perceived as deleterious. Nevertheless, as the uptake of genetic services becomes more prevalent, society may begin to recognize that negligence which leads to a failure to secure an individual's genetic project, does occasion harm to that person, even if the consequences would not universally be seen as deleterious. If society begins to look more sympathetically at frustrated genetic outcomes, the courts may feel an increasing pressure to do the same.

If the courts wanted to extend the existing boundaries of liability to recognize the harm in the novel claims, it is likely that they would currently seek to do this via incremental development of the law, rather than through massive extensions of the concept of duty of care.[1] However, Stapleton argues that where the law has developed incrementally, it has led to silly rules which threaten the dignity of the law.[2] For Stapleton, the success and dignity of tort law depends on its appearance as 'fair, sensible and focused' and she does not believe that such an approach can be achieved through incrementalism.[3]

One of the problems with the incremental development of the law is that it can generate rules which seem legally immaterial and therefore unfair.[4] In general, it seems important for the courts to maintain consistency in the law.[5] The importance of consistency in the application of negligence principles is made clear in the dicta of the majority in *Rees v Darlington Memorial Hospital NHS Trust*; a case which has been particularly prominent in this book. In *Rees*, the majority was keen to award a lump conventional sum in all cases to avoid the anomalies that might arise where damages are not awarded for the costs of raising a healthy child, but awards are made in other cases on a discretionary basis.[6] For the majority, the need to promote consistency and avoid such anomalies necessitated a wholesale approach to such claims which could be achieved by the award of a conventional sum. From the perspective that consistency is an important legal principle, this book has argued that a wholesale approach to the recognition of damage in a suite of genomic negligence claims might be adopted if the courts explicitly recognized the interest in autonomy.[7]

---

1 *Council of the Shire of Sutherland v Heyman* (1985) 157 CLR 424, Brennan J, 481, in the High Court of Australia. See Lord Keith's approval of this in *Murphy v Brentwood DC* [1991] 1 AC 398, 461.

2 J. Stapleton, 'In Restraint of Tort' in P. Birks (ed.) *Frontiers of Liability 2*, Oxford: Oxford University Press (1994), 83, 94–95.

3 Ibid. 83.

4 See in particular the rules relating to secondary victims or psychiatric injury and the distinctions made in the wrongful conception cases discussed in Chapter 5.

5 The doctrine of precedent is premised on promoting consistency. One of the earliest acknowledgements of this can be found in *Mirehouse v Rennell* (1833) 1 Clark & Finnelly 527, Parke J, 546. 'Our common law system consists of applying to new combinations of circumstances those rules of law which we derive from legal principles and judicial precedent; and for the sake of attaining uniformity, consistency and certainty, we must apply those rules, where they are not plainly unreasonable and inconvenient, to all cases which arise; and we are not at liberty to reject them, and to abandon all analogy to them, in those to which they have not been judicially applied, because we think that the rules are not as convenient and reasonable as we ourselves could have devised'. In recognizing the importance of certainty and consistency in the law, Parke J's dictum was restated in *R v Simpson* [2004] QB 118, Lord Woolf CJ, 128.

6 *Rees v Darlington Memorial Hospital NHS Trust* [2004] 1 AC 309, Lord Bingham, 317; Lord Nicholls, 319; Lord Millett, 349, all relying on the dicta of Waller LJ in the Court of Appeal.

7 The question of whether negligence law ought to value consistency over pragmatism is a complex philosophical question which is not addressed here. Relying on the courts' express pronouncements that consistency is important, it is assumed here that the courts value consistency in their approach to novel legal challenges.

Other commentators have argued that the creation of a novel 'blockbuster tort' might be the most comprehensive and flexible way of responding to novel genomic claims.[8] The argument made here is more conservative; that the recognition of a new type of damage in negligence would offer a real possibility of these novel claims being brought forward and given genuine consideration in a consistent and comprehensive fashion, which might avoid the problems that have arisen where extensions to liability are made through the ad hoc tinkering with legal rules.[9] However, controlling the boundaries of a new legal concept of harm based on interference with autonomy will not be straightforward. The principle of autonomy is not a unified entity. It is subject to particularly varied and wide-ranging interpretations. Given this, the recognition of an interest in autonomy as the basis for harm in the tort of negligence could give rise to the 'serious definitional difficulties and conceptual problems in [the] judicial development' that Mummery LJ foresaw as arsing in the judicial development of a 'blockbuster' tort to protect privacy.[10] On this basis, this book considers particular conceptions of the principle of autonomy that might enable the judiciary to gain some analytical purchase and subsequent control over the development of the principle in the context of its function as the foundation for recognition of a new form of harm.

This book argues that the recognition of an interest in autonomy is one way of responding in a comprehensive and consistent manner to novel claims which might arise from the increasing existence of genetic information and provision of genetic services. Whether this would be the best way to respond to these types of claim needs further consideration. Nevertheless, the House of Lords has recently demonstrated that it believes it to be important and legitimate for English negligence law to provide at least some protection for autonomy.[11] This cue provides the impetus for this discussion of how the law might further provide protection for this self-proclaimed important principle in the context of novel genomic negligence.

---

8 See R. Brownsword, 'An Interest in Human Dignity as the Basis for Genomic Torts' 42 (2003), *Washburn Law Journal* 143; M. Sonnenburg, 'A Preference for Non-Existence: Wrongful Life and a Proposed Tort of Genetic Malpractice' 55 (1982), *Southern California Law Review* 477.

9 J. Stapleton, 'In Restraint of Tort' in P. Birks (ed.) *Frontiers of Liability 2*, Oxford: Oxford University Press (1994) 83, 94–95.

10 *Wainwright v Home Office* [2002] QB 1334, Mummery LJ, 1351.

11 See *Rees v Darlington Memorial Hospital NHS Trust* [2004] 1 AC 309 and *Chester v Afshar* [2005] 1 AC 134.

Index